The Macat Library
世界思想宝库钥匙丛书

解析简·雅各布斯《美国大城市的死与生》

AN ANALYSIS OF JANE JACOBS'S THE DEATH AND LIFE OF GREAT AMERICAN CITIES

Martin Fuller　Ryan Moore ◎ 著
王青松 ◎ 译

上海外语教育出版社
外教社 SHANGHAI FOREIGN LANGUAGE EDUCATION PRESS
MACAT

目　录

CONTENTS

引言

要点

- 简·雅各布斯（1916—2006）是一位美国记者，致力于抨击城市重建*（一项惯常通过拆毁街区来修建高速公路的城市重建政策）和战后城市规划。
- 《美国大城市的死与生》曝光了城市规划*（一种对基础设施、交通、通讯和公共福利予以关注的城市规划过程）的失败。
- 雅各布斯立足于观察社会交往的第一手资料，提出了理解城市的另一种方法。她提出了诸种令城市更加多样、适合于步行和高密度集中的方法。

简·雅各布斯其人

简·雅各布斯，《美国大城市的死与生》（1961）一书的作者，1916年生于宾夕法尼亚州的斯克兰顿。她1934年移居纽约市，并成为一名记者，为《建筑论坛》*和其他杂志撰稿。20世纪50年代末，雅各布斯领导了一场运动，旨在将下曼哈顿（位于纽约市最大岛屿的南端，包括东村、唐人街和世界贸易中心等在内的诸多街区）从“城市重建”中拯救出来。[1]这场运动最终阻止了那些原本为了修建新道路而拆毁几个街区的计划。1961年她出版了《美国大城市的死与生》，抨击城市规划，披露二战*后城市重建的诸多败笔，当即产生了强烈反响。自那时起，许多城市规划者都接受了雅各布斯的使城市更加多元化、步行便捷化和高度集中化的思想。

雅各布斯于1968年移居多伦多，并加入了当地一个反对斯帕迪纳高速公路*的运动，这项城市重建工程计划将拆毁许多房屋、公园和小型企业。[2] 就像在纽约市一样，当地的这场运动成功地迫使高速公路建设项目被取消。雅各布斯此后一直定居多伦多。到2006年去世前，她又写了6本书，主要关于城市和经济学。20世纪70年代，她成为加拿大魁北克法语区独立运动的支持者，并在1980年出版了一本关于魁北克分裂主义问题的书。而《美国大城市的死与生》仍然是她最有影响力的著作。

《美国大城市的死与生》的主要内容

《美国大城市的死与生》挑战了城市规划和政策的主导理论。雅各布斯认为，城市规划者们摧毁大城市，是因为他们不考虑生活于其中的人们是如何生活的。她通过观察人们在街道上的交往，发展并提出了关于城市如何运行的另外一类观点。她坚信多样化、集中化和混合化是令城市伟大的良方。然而实际上，分离化和标准化却成为城市规划的核心原则。城市规划者们认为高密度和多样化会造成混乱，而雅各布斯却视这些为秩序和安全的源泉。

规划者们无视城市生活的动力，从城市理论出发杜撰出各种提议。雅各布斯认为这些不切实际的规划都是幻想，许诺一种更好的城市生活，实际上却加速了匮乏和衰退——并未考虑现实生活中数以百万的人民是如何在城市空间中居住和交往的。雅各布斯坚称，这些互动交往才是城市生活的特征。只有将这些社会动力纳入规划动议之中，城市规划才可能成功。

雅各布斯写作《美国大城市的死与生》时，正住在纽约市的格林尼治村*——曼哈顿下城一个特色鲜明的街区，她的分析都基于对当地居民和他们之间互动交往的观察。雅各布斯还与其他居民组团合作，将他们的街区从城市重建项目中拯救出来。纽约的“营造大师”罗伯特·摩西 * 计划建造一条穿越曼哈顿的高速公路，但雅各布斯和邻居们一起阻止了摩西的计划。他们的斗争非常紧迫，因为高速公路的建造可能会导致许多街区的拆迁。雅各布斯反对城市重建的斗争不只是知识层面的——它还是政治性的和个人性的。摩西的计划将会影响到雅各布斯本人，因为她就住在格林尼治村，该村被摩西谴责是“贫民窟”，但现如今，它是纽约一个音乐、艺术和文化的聚集区。

《美国大城市的死与生》为城市规划建议了几项替代性的措施。雅各布斯概括了 4 个可以在任何城市中创造多样化的条件：

- 首先，一个地区应该支持混合使用。当城市规划者分割商业的、住宅的、工业的和文化的空间时，雅各布斯则坚持这些不同的功能混合在一起会促进城市生活的改善。
- 第二，街区必须尽可能短，使它们更适宜步行，并促进交往。
- 第三，街区内应既有新建筑，又有老建筑。现代主义城市规划认为建筑总是越新越好，但老建筑能够保持街道的一种延续感。
- 最后，城市应该培育密集的人流。城市规划者坚信大量的人流是不可取的，甚至是危险的，但雅各布斯却相信密集和混合会使城市更加安全、更加赏心悦目。

《死与生》的销量超过了 25 万册，被翻译成了 6 种语言。[3] 雅各布斯的写作风格简洁易懂，折射出她的平民主义立场——而且尽管她不是一名学者，《死与生》却永远地改变了城市研究 *（一种聚焦于城市环境下的经济学、规划学、政治学、交通和社会学 * 的学术研究领域）。

《美国大城市的死与生》的学术价值

雅各布斯对城市规划的批评暴露了后者的诸多失败之处，重塑了人们理解城市的方式，同时也为城市规划带来许多新的创见。许多城市响应了她将住宅、商业、工业和文化空间相互混合的号召。尽管雅各布斯怎么说都算不上是一个规划师，但《死与生》却标志着城市研究和城市规划的一个转折点。

现实世界中关于城市空间的冲突影响了《死与生》的分析。雅各布斯和其他格林尼治村居民抨击了罗伯特·摩西的观点，尽管摩西被列为那个时代最有影响力的城市规划师。他们的运动挫败了摩西建造下曼哈顿高速公路*的计划，就此创造了历史——同时也形塑了《死与生》的分析。[4] 摩西的计划通常都围绕汽车和交通展开，而雅各布斯提醒摩西和他的同事们，人应该是首位的。摩西认为街头生活毫无价值，雅各布斯却认为街道使城市变得伟大。摩西拆毁他视作"贫民窟"的老旧社区，雅各布斯却坚持这些社区比新郊区更有价值，更具作用。《死与生》在改变城市政策与城市政治的同时，激发了相关的学术争论。

雅各布斯提出的一些建议后来被许多城市采纳。[5] 她是最早提出混合用途开发*的人之一，即各地区应把工业、商业、住宅和文化的空间相互混合，以改善城市生活。这与将不同功能区严格分割的正统理论*（亦即普遍接受的惯例）背道而驰。雅各布斯认为，新旧建筑混合能创造出更好的街道。这种对旧建筑的保护引发了一场城市历史保护运动*——一场保护、保存、修复城市中具有历史意义的建筑物、纪念碑和遗迹的运动。尽管城市规划者视大量人群的存在是一种威胁，但雅各布斯却坚持认为密集的人口是非常重要的，这种人口的混合会给所有城市带来活力和创造力。她还认为，这会使城市更加安全。混合和密度创造了监视邻里的“街道上的眼睛”。[6]雅各布斯非但不认为它是城市的混沌，反而称之为“有组织的复杂性”，会带来一种秩序感。[7]但城市规划者害怕聚集，想要分离和隔绝人群。

《死与生》背后的思想观念塑造了城市政策，影响了历史保护运动，[8]并突显了社会交往贡献出的重要的经济功能。[9]政府官员眼中的贫民窟、衰败街区和高速公路，雅各布斯却从它们身上看到了活力、潜能和一条通往城市别样景观的道路。

---

1. 安东尼·弗林特：《与摩西摔跤：简·雅各布斯如何打败纽约营造大师并改造了美国的城市》，纽约：兰登书屋，2009 年，第 3—28 页。
2. 弗林特：《与摩西摔跤》，第 182 页。
3. 斯蒂芬·沃德：“讣告：简·雅各布斯”，《独立报》，2006 年 6 月 3 日，登录日期 2015 年 8 月 29 日，http://www.independent.co.uk/news/obituaries/jane-jacobs-6099183.html。
4. 《死与生》与雅各布斯参与格林尼治村行动之间的联系在弗林特《与摩西摔跤》中有细致研究，第 95—135 页。
5. 关于雅各布斯遗产的评估，参见罗伯塔·布兰德斯·格拉茨的《哥谭之战：罗伯特·摩西和简·雅各布斯阴影下的纽约》，纽约：国家图书，2010 年，第 256—276 页。
6. 简·雅各布斯：《美国大城市的死与生》，纽约：温特吉出版社，1992 年，第 35 页。
7. 雅各布斯：《死与生》，第 429—439 页。
8. 格拉茨：《哥谭之战》，第 25—26 页。
9. 格拉茨：《哥谭之战》，第 266—268 页。

第一部分：学术渊源

1 作者生平与历史背景

要点

- 《美国大城市的死与生》对城市规划的激烈谴责产生了直接的影响，并且继续影响着有关城市生活的观念和政策。
- 简·雅各布斯对格林尼治村社会交往的观察塑造了她对城市的理解。
- 在写《死与生》的时候，雅各布斯加入到格林尼治村居民的反对运动中，一起将他们的社区从建筑师兼规划师罗伯特·摩西的重建规划中拯救出来，该规划通常都会提出拆除社区，以便建造便捷汽车的基础设施。

为何要读这部著作?

简·雅各布斯的《美国大城市的死与生》(1961)尖锐批评了罗伯特·摩西(他是一位在20世纪中叶改造了纽约及周边郊区的"营造大师")这样的建筑师，挑战了那些主导二战后美国城市规划和城市政策的核心理念。在1961年出版后，这本著作还促成了城市规划的范式转变(即彻底的重新评估)。雅各布斯发展出审视城市生活的另一种视角，即重点关注人们是如何在街道上交往的。之后，许多城市都采纳了她提出的建议，将住宅、商业、工业和文化的空间相混合。[1]

《死与生》对二战后城市规划提出了毁灭性的评判。它还预告了该书出版后的几十年里发生的许多理论和政策转变。雅各布斯对围绕汽车交通规划城市的警告已被证明是特别有预见性的。在她写作《死与生》的时候，人们清楚地看到越来越多的高架路和高速公路被建造起来，但是摩西和其他规划者对于减少交通量的承诺并没有兑现。雅各布斯看到，围绕着汽车和卡车的城市规划彻底摧毁了城市住宅区和中心商业区。[2] 随着横穿布朗克斯高速公路 * 的建造，纽约南布朗克斯的社区和居民区都在显而易见地衰落下去。雅各布斯对摩西和其他规划者的批评为城市交通方式提供了其他路径。类似这样一些源自《死与生》的思想，对当代的城市规划形式产生了至关重要的影响。[3]

> "《美国大城市的死与生》在 1961 年出版时就像一场地震，震撼了城市规划界。这本书是对规划当权派的一次正面攻击，尤其是对那些由纽约的罗伯特 · 摩西这样强大的重建官僚们组织实施的大规模城市重建项目。雅各布斯嘲笑城市重建是一个只会当即创造贫民窟的过程。"
>
> ——理查德 · T. 勒盖茨和弗雷德里克 · 斯托特:《城市读本》

作者生平

简 · 雅各布斯 1916 年出生在美国宾夕法尼亚州斯克兰顿市，1934 年和她的妹妹一起搬到了纽约市。[4] 尽管雅各布斯是一名专业记者和作家，但她既没有大学文凭也不曾受雇于大学。[5] 20 世纪 50 年代，她开始为《建筑论坛》撰写与城市和建筑有关的文章。[6] 写作《死与生》一书时，她正居住在曼哈顿下城格林尼治村的一栋翻修过的联排房屋里。[7]《死与生》取材于她对城市、规划和建筑的了解，还包括她从哈德逊街 555 号的家中观察到的情况。在《死与生》中，她写道："我所居住的那段哈德逊街上，每天都上演一幕精妙的人行道芭蕾舞。"[8]

在雅各布斯写《死与生》时，她帮助领导了一场阻止下曼哈顿高速公路 * 建设的运动。城市的独裁设计者罗伯特·摩西设想了一条将把华盛顿广场公园 * 一切为二的高速公路，而公园本是居民和被该街区的波希米亚风格吸引来的各色人等的休闲和文化娱乐中心。[9] 雅各布斯担任“阻止下曼哈顿高速公路 * 联合委员会”的主席，而且还曾在 1958 年为表示抗议而撕毁一次公开会议中速记员的笔记，并因此被捕。[10]《死与生》出版一年后，纽约市的官员们被这场运动所说服，放弃了摩西提出的计划。[11]

创作背景

第二次世界大战后，城市规划方面的主导思想包括清除贫民窟、建造高层公共住房，以及修建连接城市和郊区的高速公路。这种城市规划会将住宅、商业、工业和文化的空间分隔开来。它会为修建高速公路而把整个街区夷为平地。尽管失败的证据不断增加，但这些观念已经凝结成正统思想。城市重建运动历经数十年从设计和规划理论基础上渐渐发展成型。它的基础形成于 19 世纪，那时候，城市被视为污染的、拥挤的、普遍不受欢迎的生活地。那个时代的规划者们尝试推进乌托邦式的（也就是说，富有远见和理想化色彩的）居住和工作空间的提议和设计，以避免城市的卑污之处。[12] 他们还把城市重建视为一种改善城市的经济和社会生活的途径。但雅各布斯却认为它会导致保障城市顺利运行的一切的消亡。

与同时代的城市规划者相反，雅各布斯认为城市生活扎根于人们如何在街上相互交融。她坚信，城市应该包括混合用途的空间，将工业、商业、居住和文化功能组合成一体，而城市规划者将这些分割在不同的地区。她给大城市开的处方还包括一些易于步行的短街区，新旧建筑的交融，人口的汇聚集中。当官僚们求助于抽象的理论时，她依靠敏锐的观察来阐明她的观点："那些只对一座城市'应该'长什么样感兴趣，而对它现实运作方式不感兴趣的人会对这本书感到失望。"[13]《死与生》对城市规划者们已经毫无保留接受的观念和政策进行了无情的抨击。它猛批他们颇受质疑的项目，并最终彻底摧毁了他们的错误设想。

---

1. 当代对雅各布斯遗产的评估载于罗伯塔·布兰德斯·格拉茨：《哥谭之战：罗伯特·摩西和简·雅各布斯阴影下的纽约》，纽约：国家图书，2010 年，第 256—276 页。
2. 安东尼·弗林特：《与摩西摔跤：简·雅各布斯如何打败纽约营造大师并改造了美国的城市》，纽约：兰登书屋，2009 年，第 62—63 页。
3. 格拉茨：《哥谭之战》，第 274—276 页。
4. 弗林特：《与摩西摔跤》，第 3—4 页。
5. 弗林特：《与摩西摔跤》，第 8—9 页。
6. 弗林特：《与摩西摔跤》，第 18 页。
7. 弗林特：《与摩西摔跤》，第 99 页。
8. 简·雅各布斯：《美国大城市的死与生》，纽约：温特吉出版社，1992 年，第 50 页。
9. 弗林特：《与摩西摔跤》，第 62 页。
10. 弗林特：《与摩西摔跤》，第 xiv 页。
11. 弗林特：《与摩西摔跤》，第 158—159 页。
12. 弗林特：《与摩西摔跤》，第 20 页。
13. 雅各布斯：《死与生》，第 14 页。

2 学术背景

要点

- 简·雅各布斯挑战了两个城市规划学派：分散主义者*（自19世纪以来的一群城市理论家和规划者，他们寻求通过分散城市的人口和已建成的环境来解决城市生活的社会和环境弊病）和影响巨大的瑞士裔法国建筑师勒·柯布西耶*的弟子们。
- 尽管芝加哥学派*表达的关切点和雅各布斯相似，但雅各布斯与他们没有联系。芝加哥学派以其城市社会学（研究城市环境的社会结构）闻名。在20世纪的二三十年代，芝加哥学派基于芝加哥大学社会学系，对城市生活开展了一些具有开创意义的研究。
- 虽然《美国大城市的死与生》中的绝大部分观察来自纽约市，但雅各布斯指出美国的其他城市也被同样的问题所困扰。

著作语境

在《美国大城市的死与生》中，简·雅各布斯认为在城市规划者中存在两种思想流派。第一种是“分散主义者”，顾名思义，他们想要分离城市，分散人口。他们的主要思想来自英国城市理论家埃比尼泽·霍华德*和苏格兰城市规划师帕特里克·格迪斯爵士*，后者因提出相关观点替代他那个时代居于统治地位的城市网格规划方案而闻名遐迩。19世纪末，伦敦的环境状况使霍华德感到恐怖，于是他设想了一个“花园城市”*，在那里，人们将在乡村中创建自给自足的小城镇。雅各布斯描绘了霍华德对城市的敌意是如何影响他的城市规划的：“他不仅讨厌这座城市中乌七八糟的错误和罪恶，还憎恨城市本身，认为如此多的人聚集在一起是一种彻头彻尾的邪恶，是一种公然侮辱。他开出的拯救这些人的药方是把城市推倒彻底重来。”[1]

20 世纪初，格迪斯想把霍华德的花园城市理想扩展到整个区域，把这些人工城市分布在大城市的外围。

第二个学派的思想来源于瑞士裔法国建筑师勒·柯布西耶，他的思想在 20 世纪二三十年代指导了巴黎的重建。他理想的辐射城市 * 吁求在一个公园里耸立起一连串的摩天大楼。正如雅各布斯所写的，“他的城市就像一个奇妙的机械玩具。……它是如此有序，如此清晰，如此容易理解。它在闪念之间将一切和盘托出，就像一个绝妙的广告。”[2] 分散主义者影响了城市外围的规划，勒·柯布西耶的思想则影响了城市内部的运作。他的影响在高层办公楼和低收入住宅项目中表现得最为明显。

> “我一直在对正统的城市规划理论说些不客气的话，并会在必要的时候发表更多这样的评论。至今，这些正统观念已经成为我们习俗的一部分。它们令我们受伤，因为我们理所当然地接受了它们。”
>
> ——简·雅各布斯：《美国大城市的死与生》

学科概览

雅各布斯既批评了分散主义者，又批评了勒·柯布西耶的门徒。霍华德的花园城市认为密集的人口是一种固有的邪恶。而对雅各布斯来说，密集提供了活力和创造力的源泉。雅各布斯还蔑称，霍华德规划严整的社区愿景“即便不能说是独裁式的，但也是家长制式的。”[3] 虽然分散主义者和勒·柯布西耶派对理想社区持有不同的愿景，但是他们两者的办法都是自上而下的。勒·柯布西耶乌托邦式的理念更倾向于自由主义；正如雅各布斯形容的那样，"在他的辐射城市，想必很可能没有人会做自己兄弟的监护人。"[4] 分散主义者塑造了郊区的发展，而勒·柯布西耶的愿景在纽约市和摩西的规划中表现得最为明显。

雅各布斯本可以利用学术靠山来声援自己的思想，但奇怪的是，她从未提过以城市社会学而闻名的芝加哥学派。20 世纪二三十年代，芝加哥学派的学者们，其中包括城市社会学的奠基人罗伯特·帕克*、路易斯·沃思*和欧内斯特·伯吉斯*在芝加哥大学进行他们的研究，并因为在城市研究领域的影响而声名显赫。

雅各布斯与帕克和沃思没有交集令人惊讶，因为他们都对城市内部的社会交往感兴趣。和雅各布斯一样，帕克也有新闻业的背景（尽管是在芝加哥），并鼓励他的学生用人种学（关于人类和文化的研究）的方法研究城市生活。如果雅各布斯曾考虑过芝加哥学派的研究，特别是帕克和沃思的研究，她很可能会收获更多的启迪。

学术渊源

简·雅各布斯不是一位学者，流行的城市规划理论对她的影响仅是负面意义的。官僚主义的失败迫使雅各布斯去调查在城市中真正起作用的是什么。她将催生《死与生》的思考背后的问题归功于一位社会工作者："本书的基本思想，即试图开始理解在看似混乱的城市中复杂的社会和经济秩序，原本不是我的主意，而是威廉·柯克的，他是纽约东哈莱姆联合定居点负责人，他领着我考察东哈莱姆区，也教会我一种看待其他社区和城市中心的方法。"[5]

雅各布斯还指出，柯克请求她注意"在城市看似混乱的表象下复杂的社会和经济秩序。”[6]雅各布斯更多是通过实际观察而不是研究理论了解城市。她拒绝了规划者独裁的——极权的——本质，坚持认为社区和他们的居民对城市最了解。

因为雅各布斯的分析是建立在观察基础上的，所以她的许多例子都是来自纽约市和她居住的格林尼治村社区，这是可以理解的。然而在城市重建方面，其他的美国城市都紧随纽约的步伐，也产生了同样的负面结果。（例如芝加哥的艾森豪威尔高速公路将西郊的橡树公园彻底一劈两半，摧毁了整个社区。）雅各布斯解释说，她起初在其他城市看到许多这样的趋势：“在试图解释城市的潜在秩序时，我使用了大量来自纽约的例子，因为我住在这里。但这本书大部分的基本思想来自我起初在其他城市注意到或者听到的一切。”[7]她意识到，纽约市已经成为全美范围内其他城市的“典型”，而不断增加的败象都源自同一个有缺陷的设想。对于那些规划者们打烂了、也无力收拾的东西，她试图力挽狂澜。

---

1. 简・雅各布斯：《美国大城市的死与生》，纽约：温特吉出版社，1992 年，第 17 页。
2. 雅各布斯：《死与生》，第 23 页。
3. 雅各布斯：《死与生》，第 19 页。
4. 雅各布斯：《死与生》，第 22 页。
5. 雅各布斯：《死与生》，第 15—16 页。
6. 雅各布斯：《死与生》，第 15 页。
7. 雅各布斯：《死与生》，第 15 页。

3 主导命题

要点

- 简·雅各布斯的主要问题是城市重建究竟对城市有益还是有害。如果是有害的，她想知道城市应该寻求什么样的替代方案。
- 在雅各布斯写作《美国大城市的死与生》时，规划者们认为城市重建的好处是不言而喻的。
- 雅各布斯弃绝城市规划的思想。她坚持认为观察社会交往对于理解城市是至关重要的。

核心问题

简·雅各布斯在《美国大城市的死与生》中试图回答一个核心问题：城市重建对城市有益吗？——如果没有，那么城市应该做些什么？为了找到答案，她首先必须得解决更大的问题，即城市是如何运转的。城市规划者认为这些问题和相关回答是不言而喻的。对规划批评最猛烈的都是那些居住在被城市重建摧毁的社区里的居民。雅各布斯站在他们一边，她争辩说："规划者们在尊重和理解具体事情方面，常常显得知识储备不足，还不如没有受过专业知识训练的普通人，这些普通人与一个街区有切身的关联，习惯于使用它，而不习惯从一般或抽象的角度来考虑街区问题。"[1]

相比那些规划者——他们从来没有用雅各布斯的方式质疑城市规划的核心原则，雅各布斯相信社区能更好地理解城市。她向源于住宅区运动的城市重建提出了严苛的质问。20 世纪 50 年代，地方抗议运动反对修建横贯布朗克斯的高速公路，这条横跨纽约布朗克斯区的重要汽车公路线也是由建筑师和规划师罗伯特·摩西设计的。雅各布斯读到摩西计划将第五大道延伸并从华盛顿广场公园中间穿过的消息后，她加入了华盛顿广场公园委员会*，该委员会是由社区组织者谢莉·海耶斯*创立的。[2]《死与生》赢得读者的广泛关注，因为雅各布斯说出了人们对城市重建日益不满的情绪；它还激发了广泛争议，因为她对城市重建基本理念发起了挑战。

> "雅各布斯嘲笑城市重建只能是一个贫民窟的速成过程。她质疑广为接受的信仰条文，例如公园是有益的，拥挤的人流是有害的。相反，她暗示公园通常是危险的，拥挤的街边人行道是供孩子们玩耍的最安全的地方。"
>
> ——理查德·T. 勒盖茨和弗雷德里克·斯托特：《城市读本》

参与者

雅各布斯既抨击分散主义者的"摧毁城市的思想"，又抨击勒·柯布西耶的城市规划。[3]

分散主义者的思想肇始于埃比尼泽·霍华德的花园城市模式，他们赞成一种源于厌恶城市及其密集人口的规划路径。在雅各布斯时代，《纽约客》*杂志的建筑评论家刘易斯·芒福德*是分散派最有影响力的信徒。芒福德写了一本书叫《城市文化》，雅各布斯嘲笑它是"一个病态的、有偏见的疾病目录。"[4]作为回应，他为《死与生》撰写了一篇评论文章，题为《雅各布斯妈妈的居家良方》。[5]芒福德的评论认可了《死与生》的重要意义及其批判的新颖性，但他用一种高傲的甚至是沙文主义的语调，把雅各布斯的做法称作"赤裸裸的无知"，并将其描述为"一种理智感伤情绪的混合物、一种成熟判断和女生咆哮的混合物。”[6]

罗伯特·摩西从未对《死与生》发表过公开评论，但是他把该书的出版商兰登书屋寄给他的样书退了回去，并附上一封致该公司联合创始人的措辞严厉的信。“我把寄给我的书退还给你，”摩西写道，“它除了言辞过激和不够严谨以外，还造谣诽谤……把这垃圾卖给别人吧。”[7]《死与生》显然触动了这位自认为是纽约营造大师的男士的神经。

当代论战

雅各布斯写作《死与生》时，大体而言，规划者们还并不质疑城市重建的明智——但是她关于什么使城市美好的思考挑战了他们的假设。哥伦比亚大学教授萨斯基亚·萨森 * 认为，理论家曾把城市视为“了解更庞大过程的一面透镜，而它的这个角色在 20 世纪 50 年代已经失去了。”[8]但适逢其时，随着雅各布斯的思想开始获得越来越多的接受，一些规划者开始采纳她的建议。总之，“雅各布斯的思想对城市学家和规划者思考城市生活的方式产生了巨大的影响。”[9]随着《死与生》的出版，规划者和建筑师开始更加关注城市中的社会交往动力学，城市重建开始受到更加密切的关注。[10]就此而言，雅各布斯确实领先于她的时代。

同样，研究城市和城市发展理论的学者们采纳了更多雅各布斯的观点。《死与生》后来成为著名的“新城市主义” * 学派的奠基性文献，[11]学者们开始从城市中产阶级化 *（富有的专业人士迁居到城市中心区的过程）角度重新反思雅各布斯的著作。她对混合用途开发、老旧建筑和充满活力的街道生活的看法影响了城市中产阶级化的进程。而且，随着城市中产阶级取代工人阶级家庭，雅各布斯对权力和不平等的忽视受到了密切关注；[12] 与此类似，她也没能考虑到种族、族裔、阶级和性别这些因素也影响着人们对城市的体验。然而，她在规划界的对手们甚至从未把人放在首要位置进行考虑，而从此以后，权力和不平等这些关键问题都成为城市研究的核心。

---

1. 简·雅各布斯：《美国大城市的死与生》，纽约：温特吉出版社，1992 年，第 441 页。
2. 安东尼·弗林特：《与摩西摔跤：简·雅各布斯如何挑战纽约营造大师并改造了美国的城市》，纽约：兰登书屋，2009 年，第 75 页。
3. 雅各布斯：《死与生》，第 17 页。
4. 雅各布斯：《死与生》，第 20 页。
5. 刘易斯·芒福德："天际线：'雅各布斯妈妈的居家良方'"，《纽约客》，1962 年 12 月 1 日。
6. 肯尼斯·基德："简·雅各布斯的批评者真的有价值吗？"，《多伦多星报》，2011 年 11 月 25 日，登录日期：2015 年 8 月 31 日，http://www.thestar.com/news/insight/2011/11/25/did_jane_jacobs_critics_have_a_point_after_all.html。
7. 弗林特：《与摩西摔跤》，第 125 页。
8. 萨斯基亚·萨森："简·雅各布斯在全球城市、地区、社会实践中看到了什么？"，载《简·雅各布斯的城市智慧》，索尼娅·赫特和戴安娜·扎姆编，纽约：劳特利奇，2012 年，第 84 页。
9. 马克·戈迪纳、雷·哈钦森和迈克尔·T. 瑞安：《新城市社会学》，第 5 版，科罗拉多州博尔德：韦斯特维尔出版社，第 328 页。
10. 罗伯塔·布兰德斯·格拉茨：《哥谭之战：罗伯特·摩西和简·雅各布斯阴影下的纽约》，纽约：国家图书，2010 年。
11. 彼得·卡茨：《新城市主义：走向社区建筑》，纽约：麦格劳–希尔教育集团，1993 年。
12. 莎伦·佐金：改变权利景观：富足和对真实性的渴望，《国际城市和区域研究杂志》第 33 卷，2009 年第 2 期，第 548—549 页。

4 作者贡献

要点

- 简·雅各布斯旨在通过调查城市规划者所忽视的问题来展示城市是如何运转的。
- 城市规划者相信城市有很多问题，需要重建。来自芝加哥大学社会学系的芝加哥学派的社会学家们开展研究，也认为城市是不尽如人意的。
- 雅各布斯探究城市是如何运转的，是因为她可以看出城市规划是在破坏而非重建城市。

作者目标

在《美国大城市的死与生》中，简·雅各布斯试图理解并解释是什么使城市有益于居住其中的人群。这项工作提出了一些城市规划者没能提出的问题，而这很大程度上是因为他们认为城市是不适宜居住的地方。战后的政策支持人口迁移到新的郊区，而规划者则提倡摧毁城市街区的城市重建模式。

规划者对城市的看法很抽象，几乎忽略城市居民的生活和需求。相反，他们关心的是建筑物的高度，绿色空间的创建，以及有效的交通流。雅各布斯将焦点重新拉回到人们如何使用城市空间方面，并重新定义了他们的问题。她在《死与生》开头写道："我将主要讲述一些平常的、普通的事情。"[1] 她随后列举了她所关注的问题，包括：

- 城市街道的安全

- 城市公园的质量
- 贫民窟的状况，以及它们为何尽管遭到强烈反对却时时会重生
- 城市中心区关切点的转移
- 城市街区的性质和功能[2]

雅各布斯成功地扭转了关于城市的辩论要点。她的发现以观察为基础，因此得出的结论与城市规划者们大不相同。

> “简·雅各布斯的思想影响了城市主义者，因为她抓住了城市文化的核心和灵魂。她的重要性在于使我们相信城市文化依靠个人交往和公共空间之间的关系。”
>
> ——马克·戈迪纳、雷·哈钦森和迈克尔·T. 瑞安：《新城市社会学》

研究方法

雅各布斯取得了新突破，因为她关注的是人们如何互动交往和使用城市空间。她的方法与城市规划的抽象理论完全相反。城市规划者视城市生活毫无价值可言，他们谴责整个社区为“贫民窟”，并开始摧毁它们。对他们来说，美好的生活只在远离美国都市的地方存在，在小城镇和郊区。但雅各布斯揭示了城市是如何充满活力、创新和多样性的。简言之，“对雅各布斯来说，积极的城市生活决不可能被规划，因为是人发明了空间的用途。”[3]

《死与生》的创造性在于雅各布斯研究城市生活的方法。她关注邻居们的日常生活是如何进行的，观察到一个城市的社会秩序“基本上是由一个错综复杂的、几乎无意识的网络维持的，这张含

有控制权和标准的网络是人们之间自愿订立的，并由人们自愿强制执行。”[4] 换句话说，城市街道上的社会交往创造出一种秩序形式，它不是规划人员和建筑师所能设计的。

相比之下，她看到高层住宅项目是如何变得乏味和危险的。关于这样的地方，她写道："在常规的、随机的强制文明秩序已经崩溃的地方，警察再多也不管用。”[5] 她还注意到战后郊区是如何变得千篇一律、单调乏味的。这是一种超前于时代，在未来几十年内都不会被接受的观点。雅各布斯聚焦于城市生活中的人类活动范围，运用原始的观察式方法，得出了与城市规划非人格化模型截然相反的新颖结论。

时代贡献

几乎没有学者曾通过日常的社会交往来研究城市生活，因此《死与生》的原创性来自雅各布斯的街头分析，这一点彻底颠覆了城市规划师们自上而下的研究方法。尽管雅各布斯没有与以城市社会学闻名的芝加哥学派有过联系，但学派的成员们也提出过类似的疑问。芝加哥学派更多地关注城市亚文化，而不是已建成的城市环境，其中芝加哥大学社会学家路易斯·沃思更加关切作为一种生活方式的**城市主义**，而不是作为一种物质形变的**城市化**。对于沃思来说，城市主义的三个特征是总体人口规模、密度和异质性（这是指事物内部在诸如特征、用途和人口等方面的差别）。[6]

《死与生》被证明是具有原创性的，因为雅各布斯不仅研究街道层面的城市，而且得出了积极的结论。规划者和人种学家（指研究一个地方的居民的学者）都认定城市生活是不受欢迎的。甚至芝加哥学派的城市社会学家也倾向于把城市主义看作是一种社会弊病。而且在二战后的那些年里，规划者鼓吹减少城市人口规模，鼓励去郊区生活。雅各布斯的目标在她的时代显得引人注目，是想要展现城市究竟是如何同时培养创造力和社区的。像罗伯特·摩西这样的大人物把城市社区看作是急需拆除重建的混乱的贫民窟。雅各布斯争辩说，尽管城市居民很穷，但这些社区却是交往的动力的中心。城市规划非但没能重建它们，恰恰相反，却使得美国城市中数不清的角落变得更加平淡无奇、荒凉和疏离。

---

1. 简·雅各布斯：《美国大城市的死与生》，纽约：温特吉出版社，1992 年，第 3—4 页。
2. 雅各布斯：《死与生》，第 3—4 页。
3. 马克·戈迪纳、雷·哈钦森和迈克尔·T. 瑞安：《新城市社会学》，第 5 版，科罗拉多州博尔德：韦斯特维尔出版社，第 327 页。
4. 雅各布斯：《死与生》，第 32 页。
5. 雅各布斯：《死与生》，第 32 页。
6. 路易斯·沃思：《城市主义作为一种生活方式》，《美国社会学杂志》第 44 卷，1938 年 7 月第 1 期，第 1—24 页。

第二部分：学术思想

5 思想主脉

要点

- 简·雅各布斯的主要论点是，城市规划者之所以毁灭城市，是因为他们没有考虑到城市社会交往的动力学——不同背景下一个城市被各种不同社会交往行为所规定的方式。
- 雅各布斯反对城市规划中的隔离的、标准化的手法。她坚持多样性、集中性和混杂性会使城市更伟大。
- 雅各布斯《美国大城市的死与生》的语言是直白的口语体。这反映了她对抽象城市理论的平民主义批判立场。

核心主题

《美国大城市的死与生》认为城市规划摧毁了城市生活的活力。简·雅各布斯坚持认为规划者失败了，因为他们并不理解是社会互动交往创造出伟大的城市。虽然他们把城市重建作为进步的一种形式呈现给公众，但雅各布斯认为这实际上让城市变得比以前更糟糕、更危险。在 1961 年的书中，她评估了城市更新后的城市状况："低收入住宅区成了青少年犯罪、蓄意破坏和普遍的社会绝望情绪的中心，它们原本是要取代贫民窟，但现在比贫民窟更糟糕……人行道不知道起自何处，通往何方，也没有一个漫步者。高速公路令大城市元气大伤。这不是城市的重建，这是对城市的洗劫。"[1]

雅各布斯在《死与生》中阐发了 4 个关键主题：

- 城市重建不是在重建城市，实际上是在摧毁城市。
- 规划者忽视了城市社会生活的日常现实。

- 人们在街道上的交往让城市“伟大”。
- 大城市是建立在多样性、集中性和混杂性的基础上，而不是分离化和标准化。

由于规划者提出的建议和政策主要来自城市理论，所以他们忽略了对城市中生活的微观社会动力学——个人与个人之间的交往——的考虑。雅各布斯将这视作为一种未来主义幻想，允诺给人一种更现代化的生活方式，却完全不出意外地导致失败和痛苦：“作为伪科学的城市规划及其伙伴城市设计艺术，还没有和各种愿景、似曾相识的迷信、过度简化，以及各种符号带来的似是而非的舒适感断绝联系，还没有开始踏上探索现实世界的冒险历程。”[2]

官僚们（整天坐在办公室里的城市官员）非但不利用他们的观察权力，反而眼睛死盯着建筑物和高速公路，根本不把受这些建筑物影响的人纳入考虑范围。

> “这本书是对当下城市规划和重建的抨击。更主要的，它也是试图引入一些城市规划和重建的新原则，这些原则与现在所教授的一切——从各个流派到建筑与规划，到周末副刊和女性杂志——都不同，甚至截然相反。”
>
> ——简·雅各布斯：《美国大城市的死与生》

思想探究

雅各布斯的严厉批评反映出一种与众不同的视角，这种视角得自于她在纽约市格林尼治村街区的亲身经历。她坚持认为，任何城市的生活都存在于数以百万计的人口的日常行动和交往中——城市规划建议只有在将这些社会动力纳入考虑范畴的情况下才能成功。

雅各布斯写道："只能因为且当每个人都参与了创造时，城市才有能力为每个人提供一些东西。"[3]她坚持，创造伟大城市的是参与交往互动的整个社区的居民，而不是一支专业建筑师组成的团队。

然而，这些专业人士和公众官员们为分隔和标准化到处游说，将商业、住宅、工业和文化的空间分隔开来。这导致了城市范围之外同质化的、刻板的郊区，以及城市内部千篇一律的、孤立的住宅区。相比之下，雅各布斯坚持认为多样性、集中性和混杂性使城市更加伟大。"要想理解城市，"雅各布斯认为，"我们就必须坦率地面对多种用途的组合或混杂才是必不可少的现象，而不是各自分割的用途"。[4]和当时流行的观点相反，雅各布斯主张城市应该合并空间，混合用途。这一想法在当代城市规划中具有相当大的影响力，[5]而她与之搏斗的惯例也在今天被视作与创建充满活力的城市不相容而被摒弃。

多样性、集中性和混杂性也适用于城市中的居民以及他们的交往方式（雅各布斯写道："人们集中在具有一定规模和密度的城市里，可以被看作是一个积极的因素。"[6]）。密度与多样性有着紧密的关联，这是雅各布斯重视的城市生活的另一个方面，但是她那个时代的城市规划者倾向于害怕密集的人群。雅各布斯认为，不同类型的人之间的交往在创造城市活力方面发挥着重要作用，因为她支持"巨大而丰富的差异性和可能性，许多这样的差异性是独一无二、不可预测的，而正因如此它们才更有价值。"[7]

语言表述

简·雅各布斯对那些用远离现实生活的抽象概念来设计城市的所谓专家们很不耐烦。她相信城市代表着普通人和平凡行为的聚集。同样地，她写作《死与生》用的是朴素、简洁的散体文，避免了城市规划和理论中的行话和冷冰冰的语言。雅各布斯简练、直截、率真的文风从《死与生》的开篇第一句就跃入了眼帘："本书是对当下城市规划和重建的抨击。"[8] 雅各布斯直言不讳的笔调与她平民主义的批判内容是一种完美结合。

雅各布斯并不是作为一个学者跻身城市和城市规划研究中的，而是作为一名记者和纽约居民。这一点，再加上她的观察的平实性本质，赋予《死与生》一定程度的亲切感，让它仿佛是一篇写得很好的杂志文章。20 世纪 40 年代初，雅各布斯开始当记者，后来又为《建筑论坛》杂志撰稿。正如安东尼·弗林特在他的《与摩西摔跤》一书中所写："雅各布斯在新闻和杂志出版的方方面面都颇有天赋——一个执着于细节的人，一名在正规写作风格和语法问题上的专家，条理性极强，而且很擅长讲故事。"[9] 雅各布斯把她作为记者和作家的非凡才华运用到了《死与生》中，导致了一场摧毁性的批评运动，阻止了一个规划中的高速公路项目，并促成了一种具有远见的城市视野。

---

1. 简·雅各布斯：《美国大城市的死与生》，纽约：温特吉出版社，1992 年，第 4 页。
2. 雅各布斯：《死与生》，第 13 页。
3. 雅各布斯：《死与生》，第 238 页。
4. 雅各布斯，《死与生》，第 144 页。
5. 罗伯塔·布兰德斯·格拉茨：《哥谭之战：罗伯特·摩西和简·雅各布斯阴影下

的纽约》，纽约：国家图书，2010年，第256—276页。

6. 雅各布斯：《死与生》，第220—221页。
7. 雅各布斯：《死与生》，第220—221页。
8. 雅各布斯：《死与生》，第3页。
9. 安东尼·弗林特：《与摩西摔跤：简·雅各布斯如何挑战纽约营造大师并改变了美国的城市》，纽约：兰登书屋，2009年，第10页。

6 思想支脉

要点

- 简·雅各布斯提出一些建议来补充她对城市规划的批评。这些建议包括混合用途开发（工业、商业和文化活动发生在一起的空间），短街区，老建筑和密集化住宅。
- 当代城市规划者接受了雅各布斯的批评，也采纳了她许多的建议。
- 雅各布斯倡导的混合用途空间对当代城市规划的影响最大。

其他思想

简·雅各布斯《美国大城市的死与生》主要关注的是城市重建的影响。在解释城市重建失败的原因时，雅各布斯调查了问题的另一面：是什么使城市运转？是什么曾使它们伟大？城市规划者认为多样性是混乱和失序的根源，而雅各布斯反驳说，多样性是力量和活力的源泉。

虽然《死与生》对城市重建发出了尖锐的批评，但它也提供了明智的选择；在著作的第二部分，雅各布斯概述了创造城市多样性的 4 个条件：

- 首先，每个地区应该包括一个以上的基本功能，从而创造混合用途的空间。
- 第二，大多数街区应该是短的，这样才有更多的机会转弯。
- 第三，当一个地区的建筑物在楼龄和状况上存在差别时，它才是最好的。
- 第四，应该有足够密集的人口集中度。

雅各布斯认为，所有 4 个条件都必须具备才能促进多样性，因此，缺少了其中任何一个条件，其他条件就会受到限制。“因为种种不同原因，不同地区的潜力也各不相同，”她指出，“但只要这 4 个条件得到了发展，那么一个城市的地区无论位于何处，应该都能发挥其最大的潜力。”[1] 并非所有的城市都一样，但雅各布斯相信它们都能从这 4 个条件中获益。

> “在看似混乱的表象下，老城还能成功地运行，是因为它有一种维护街道安全和城市自由的绝妙秩序。这是一种复杂的秩序，其精华是人行道用途的精细复杂性所带来的一连串的目光。”
>
> ——简·雅各布斯：《美国大城市的死与生》

思想探究

城市规划者更倾向于将住宅、商业、工业和文化区域分隔开来。而雅各布斯主张混合用途开发能确保一天中不同时间段里人群的循环分流。多种用途的空间吸引了“人流的到来，他们按照各自不同的日程出门，因为不同的目的来到此地，但是他们都能使用许多共用的设施。”[2] 当街道上全天都有人的时候，社区会变得更安全，当地的商业也会更繁荣。

短街区对于促进多样性也有类似的重要性。城市规划者们创造出漫长的街道和分散的社区导致了孤立、不景气和危险的不断绵延。相反，雅各布斯提出，“繁忙的街道和短街区是很有价值的，因为它们可以让城市社区的使用者处于一种精细复杂、交叉使用的结构。”短街区能促使为了不同目的使用街道的行人在街道上的移动与混合。[3]

雅各布斯关于城市需要老建筑的主张，既违背了城市规划的常理，也违背了战后现代主义的理想，后者在许多艺术和文化领域里赞扬新的，贬低旧的。官僚主义者和他们的建筑师把拥有老旧建筑的社区看作是祸根。但当他们推进拆除重建进程，并将其看作是一种进步时，雅各布斯反其道而行之："如果一个城市只有新建筑，那么能够在这里生存下去的企业自然而然只有那些能够承受高昂建筑成本的企业。"[4] 她认为，保护老建筑是保护能在其中运转的本地企业的关键，否则这些企业会被新建筑的高昂日常开销所淘汰。

雅各布斯提出的多样性的第四个条件是人口密集程度。她又一次违背了城市规划机构的主流观念。毕竟，规划师们鄙视和害怕密集的人口。但是雅各布斯认为，人口密度是使城市变得有趣和充满活力最重要的因素之一，她还坚称大量的人口会给城市带来独一无二的特性："密集的人口也应该作为一种财富来享受，他们的存在值得庆祝：在一个需要繁荣城市生活的地方，应该提高地区的人口密度，此外，还应该致力于创建一种非常活跃的公共街道生活，并在经济和视觉两方面竭尽所能地鼓励和支持多样性。"[5]

被忽视之处

自那时起，许多规划者都采用了雅各布斯的方法来促进多样性。她的想法曾经招致过规划机构的鄙视，但现在实际上已成为主流规划的一部分。结果，它们与社区组织的原始性关联（与精英专家相对立）现在被遗忘了。雅各布斯认为，社区对城市最了解，但一旦规划者们开始采用她的思想，这种洞察力就消散了。简而言之，"雅各布斯对社区组织性的影响总是被忽视了……雅各布斯实干家式的工作向全国人民表明，他们能够与城市重建的推土机进行斗争——并夺取胜利。”[6] 当代城市主义者称颂雅各布斯在改变了城市该如何规划的争论中的贡献。然而，他们却忽视了她的思想是如何在一个投身于政治斗争的更大规模的社区中产生的。

《死与生》另一个被忽视的方面涉及妇女与城市之间的关系。该书写于 1961 年，雅各布斯的分析比第二波女性主义 *（20 世纪 60 年代末开始的妇女社会运动，延续着早期争取非歧视、生殖权和社会平等的斗争）的出现早了好几年。政治学家马歇尔·伯曼 * 认为，“她让读者们觉得，女性对于城市生活（每一条街，每一个日子）要远比规划建造城市的男人们更加了解。”[7] 通过审视日复一日发生的平凡活动，雅各布斯展示了自己对城市生活的视角更加契合自己的女性身份。在伯曼看来，“自［社会改革家］简·亚当斯 * 以来”，雅各布斯“第一次充分阐述了女性对于城市的观点。”[8]

主流地理学家和规划师经常绕过雅各布斯批评中的女性主义，因为它的出现早于第二次女性主义浪潮。尽管如此，《死与生》在一个至关重要的社会和政治话题上体现了女性主义视角特质，见证了一个意志坚定、富有远见的女性面对一个根深蒂固的男权中心所能达到的成就。

---

1. 简·雅各布斯：《美国大城市的死与生》，纽约：温特吉出版社，1992 年，第 151 页。
2. 雅各布斯：《死与生》，第 152 页。
3. 雅各布斯：《死与生》，第 186 页。

4. 雅各布斯:《死与生》，第 187 页。
5. 雅各布斯:《死与生》，第 221 页。
6. 皮特·德瑞尔,《简·雅各布斯的激进遗产》,《国家住房研究所》146 期，2006 年夏，登录日期 2015 年 9 月 5 日，http://www.nhi.org/online/issues/146/janejacobslegacy.html。
7. 马歇尔·伯曼:《一切坚固的东西都烟消云散了：现代性经验》，纽约：企鹅图书，1982 年，第 322 页。
8. 伯曼:《一切坚固的东西都烟消云散了》，第 322 页。

7 历史成就

要点

- 简·雅各布斯揭露并粉碎了第二次世界大战后城市重建中改造政策的基本前提。
- 雅各布斯的论点之所以占上风，是因为她表达了对城市重建的普遍不满。
- 因为没有兼顾考虑经济因素和政治因素，雅各布斯没有给批判城市中产阶级化（富裕的专业人士迁移到重建的城市社区，从而抬高房价挤走现有居民的过程）留下任何余地。

观点评价

简·雅各布斯的《美国大城市的死与生》现在被认为是城市研究的经典之作，经常作为关键文献在城市社会学*、地理学和建筑学领域被引用。

雅各布斯的建议也经常作为政策付诸实施，正如最近出版的《新城市社会学》一书所说："雅各布斯的观点对城市学家和规划者思考城市生活的方式产生了强烈影响。地方政府鼓励公园利用、街头节庆、临时封锁社区道路，以及容忍人行道上的小贩。"[1] 的确，在数量众多的学科领域里，很少有几本书能对理论和实践做出如此多的贡献。《死与生》的出版标志着城市研究的一个转折点。

然而，并非《死与生》的所有思想都被证明是可接受的或成功的。《新城市社会学》的作者们写道，"她的一些追随者主张移除公寓建筑的电梯以方便邻里间的交往互动，但其结果对这些建筑物的居民们来说却是灾难性的。”[2] 妨碍雅各布斯思想落实的一个关键因素是犯罪。所以，在许多城市，“利用雅各布斯思想振兴市中心区的努力失败了，其原因是郊区居民对城市犯罪的恐惧。”[3] 最后，《死与生》中弥漫着一种不再与城市生活相契合的强烈感伤和怀旧之情。总之，“雅各布斯关于社区的思想也可能过时了。许多城市居民是与并不住在附近的朋友、亲戚交往……青少年可能更喜欢去找自己朋友圈里的朋友们玩，而不是在街上交朋友。”[4]

> “伟大的城市理论家简·雅各布斯不是一名受过学术训练的经济学家，但她的增长理论为这一领域做出了不可磨灭的贡献。在她看来，是新的工作类型和新的做事方式推动了大规模经济扩张。大多数经济学家将动力定位在大公司、企业家和民族国家身上，雅各布斯则认为大城市才是创新背后首要的原动力。”
>
> ——理查德·弗罗里达：《创意阶级的崛起》

当时的成就

雅各布斯写作《死与生》的时代适逢社区开始挑战城市重建之时。这些社会环境促成了该书的成功。20 世纪 50 年代，社区运动在纽约的布朗克斯区开始兴起，反对修建一条主干道——横穿布朗克斯高速公路。1961 年，在居民们与建筑师兼规划师罗伯特·摩西的下曼哈顿城市重建计划作斗争时，《死与生》出现了。它阐述了一种新的城市思考方式：“它不仅是新一代的规划师和建筑师的灵感、指南和圣经，也是普通公民们的灵感、指南和圣经。”[5] 这些外部政治冲突事件令《死与生》成为当时最急需的书籍之一。

《死与生》是 20 世纪 60 年代初对社会问题和公共政策造成强

烈影响的几本书中的第一本。紧随其后出版、引起公众关注的其他书籍包括海洋生物学家和环保主义者蕾切尔·卡森*的《寂静的春天》(1962)、社会活动家迈克尔·哈林顿*的《另一个美国》(1962)、女权主义作家贝蒂·弗里丹*的《女性的奥秘》(1963)、活动家和消费者权益保护者拉尔夫·纳达尔*的《任何速度都不安全》(1965)。卡森的《寂静的春天》揭露了杀虫剂的滥用，从而帮助推动了环保运动。[6]哈林顿的《另一个美国》为反贫困斗争提供了催化剂。[7]《女性的奥秘》成为第二波女权主义运动的奠基文献；弗里丹于1966年成立全国妇女组织。[8]纳达尔的《任何速度都不安全》暴露了汽车工业的安全措施缺失（将美国汽车制造商雪佛兰的科维尔车型作为关注焦点），并为消费者保护提供了动力。[9]

《死与生》也为预示着城市中产阶级化的"回归城市"运动提供了理论依据。从20世纪50年代开始，纽约的中产阶级专业人士开始返迁到布鲁克林区以前的工人阶级社区，如布鲁克林高地、公园坡和波伦山，并翻新那里古典的老房子。[10]雅各布斯的书出版后在他们当中产生了很大影响："大多数受到简·雅各布斯的《美国大城市的死与生》一书启发搬迁到布鲁克林的中产阶级狂热人士，都念叨她富有感伤情绪的'街头芭蕾'，或者她那些悄悄坠入浪漫的怀旧情怀的文字。"[11]但《死与生》没有料想到的一个效果是它对城市中产阶级化的刺激作用——当代评论家都责怪说，城市中产阶级化将工人阶级、穷人、有色人种和族裔群落驱离了城市社区。

局限性

《死与生》出版后的几十年里，学者们发现了其中的至少两个主要缺点。首先，雅各布斯没有注意到推动城市发展的经济因素。

尽管她不相信罗伯特·摩西所代表的政府权力，但是资本家开发商也逃脱了她的批评。所以，在“雅各布斯和其他［格林尼治村社区的居民们］忙于抵抗摩西的大规模公共工程计划时，他们未能抵挡住私人房地产开发商的更加悄无声息的、一点一滴的入侵。整个20世纪50年代，这些开发商在村里遍地拆除旧建筑，建造新的高层建筑。”[12] 虽然《死与生》对城市重建提出了摧毁性的评估，但雅各布斯对资本主义发展的疏忽坐实了她与城市中产阶级化的共谋。

第二个缺点与雅各布斯对大城市的怀旧式的理想有关，这个理想认为老旧社区要比其郊区的继任者们更亲密、更脚踏实地。这源于雅各布斯想象中族裔社区的真实性。在《死与生》出版后，城市中产阶级化的专业人士返回城市寻找这种真实性。简单地说，“雅各布斯没有认识到她自己观点越来越大的影响力，没能看见像她这样的家庭正逐渐搬到西村19世纪的房子里，因为他们欣赏该地区小商店和鹅卵石街道的魅力。”[13] 换句话说，雅各布斯把老城市和城市社区理想化了，其方式后来被年轻的白领专业人士所效仿，促使像芝加哥和旧金山这样的美国城市走向中产阶级化。城市中产阶级化的吊诡之处在于，它将专业人士描绘为真实性化身的族裔和艺术连根拔除了。

---

1. 马克·戈迪纳、雷·哈钦森和迈克尔·T. 瑞安：《新城市社会学》，第5版，科罗拉多州博尔德：韦斯特维尔出版社，2015年，第328页。
2. 戈迪纳、哈钦森和瑞安：《新城市社会学》，第328页。
3. 戈迪纳、哈钦森和瑞安：《新城市社会学》，第328页。

4. 戈迪纳、哈钦森和瑞安:《新城市社会学》,第 328 页。
5. 安东尼·弗林特:《与摩西摔跤:简·雅各布斯如何挑战纽约营造大师并改变了美国的城市》,纽约:兰登书屋,2009 年,第 129 页。
6. 蕾切尔·卡森:《寂静的春天》,波士顿:霍顿·米夫林公司,1962 年。
7. 迈克尔·哈林顿:《另一个美国:美国的贫困》,纽约:塔奇斯通,1962 年。
8. 贝蒂·弗里丹:《女性的奥秘》,纽约:W.W. 诺顿,1963 年。
9. 拉尔夫·纳达尔:《任何速度都不安全:美国汽车设计中的危险》,纽约:格罗斯曼,1965 年。
10. 苏莱曼·奥斯曼:《褐砂石建筑布鲁克林的发明:城市中产阶级化与战后纽约真实性的找寻》,纽约:牛津大学出版社,2011 年。
11. 奥斯曼:《褐砂石建筑布鲁克林的发明》,第 169 页。
12. 约翰·斯特劳斯堡:《村:格林尼治村的历史》,纽约:哈珀·柯林斯,2013 年,第 341 页。
13. 莎伦·佐金:《裸城:真实城市空间的死与生》,纽约:牛津大学出版社,2010 年,第 18 页。

8 著作地位

要点

- 简·雅各布斯兴趣广泛，而对城市及其经济的专门研究贯穿她的一生。
- 尽管她还写了 6 本书，但《美国大城市的死与生》一直是她最著名的作品。
- 《死与生》确立了雅各布斯作为一位城市和城市生活研究领域国际性权威的声誉。

定位

《美国大城市的死与生》堪称是简·雅各布斯最著名的作品。在更早的 1958 年发表于《财富》杂志的《市中心是人民的》的文章中，她曾概述过这本书的一些主旨。雅各布斯在哈佛大学设计学院研究生院举办的城市设计会议上发表过一次演讲，《财富》杂志的作者兼编辑威廉·H. 怀特 * 请她把这次演讲写成了一篇文章。在《市中心是人民的》一文中，雅各布斯提出了她对城市重建的主要批评："这些项目不会振兴市中心；它们将抑制市中心……它们将是稳定的、对称的、秩序井然的，它们将是干净的、吸引人的、纪念碑式的，它们将拥有一块保存良好的、庄重的墓地。"[1]《财富》杂志的这篇文章阐述了她反对城市重建的部分观点，但她还没有明确是什么在推动城市运转。

《死与生》是雅各布斯 7 本书中的第一本，7 本书中的大部分都聚焦于城市与城市发展。在她积极参与纽约行动的几年中，她就开始写第二本书《城市经济学》，并在搬到多伦多后不久完成了。在“城市珍贵的低效和不切实际”这一章中，雅各布斯比较了英国城市曼彻斯特和伯明翰的经济状况。当曼彻斯特在“淘汰的城市”边缘摇摇欲坠时，伯明翰的“零散低效的小工业不断增加新的工作，且分离出新的组织，其中一些变得非常庞大，但在总就业人数和总产量方面仍然不及小工业。”《城市经济学》继续发展了她在《死与生》中得出的一个中心结论：多样性对于城市生活至关重要，而大规模的项目适得其反。[2]

> “[她的其他]书没有哪一本像《美国大城市的死与生》那样洛阳纸贵，而当这些问题不可避免地回溯到她置身格林尼治村的波希米亚人中间，参加纽约之战的那些时日时，雅各布斯开始恼火起来：仿佛她是个摇滚明星，一直被要求唱老歌。”
>
> ——安东尼·弗林特：《与摩西摔跤：简·雅各布斯如何挑战纽约营造大师并改变了美国的城市》

整合

雅各布斯余下的著作探讨了一连串其他问题，往往偏离对城市及其经济的分析。她的第三本书，1980 年出版的《分裂主义问题》支持说法语的魁北克地区脱离加拿大独立。[3] 虽然这本书把她的关注点从城市中分散出来，但它继承了从《死与生》开始的广受欢迎的自治倡议。她的下一本书《城市和国家财富》重新讨论城市经济学的主题，但第五本书《生存系统》探讨了新的道德和价值观问题。[4]《生存系统》研究了雅各布斯称之为“守护者综合征”和“商业综合征”背后不同的道德戒律。在《经济学的本质》中她回归到经济学研究，但她最后一本书《即将来临的黑暗时代》对北美社会的衰败发出了世界末日般的警告。[5]

尽管雅各布斯对城市和经济学保持着毕生的兴趣，但她的书并没有集中在一个单一的、连贯的思想上。相反，她探讨了广泛领域里的各种问题和争议。即便如此，她的《死与生》仍然是她最有影响力的著作。雅各布斯认为自己是一位兴趣广泛的学者，公众则总把她和城市以及20世纪60年代的反对城市重建联系在一起。

意义

她的其他书基本上都被遗忘了，但《死与生》已经被证明有持久的影响。它不但确立了雅各布斯作为城市研究一流专家的声誉，而且信息丰富，又予人启迪。安东尼·弗林特在他阐述雅各布斯为一种新的城市再生而战斗的书《与摩西摔跤》中写道："这本书对新一代城市社会活动家、学生——他们把她看作一位人民英雄——以及城市规划者的影响是无可否认的。全美各城市的活动家都以雅各布斯为榜样，充当地方政府的监督员，要求对一切问题发表意见，从街角的废纸篓到摩天大楼投下的阴影。"[6]

《死与生》的持久影响至今已持续了半个多世纪，而且还在与日俱增。雅各布斯在书中预言了20世纪60年代后城市将经历怎样的变化。例如白领、中产阶级专业人士现在发现城市是很理想的居住地，其原因与雅各布斯最初提出的许多理由一样。她的思想也在肇始于20世纪80年代并不断铺开的后现代建筑*形式中得到实现（在建筑学中，"后现代"指的是一场试图摆脱战后主流建筑风格，特别是被称为"现代主义"的建筑风格的运动）。正如城市理论家大卫·哈维*所写的那样："因此，与战后初期的情形相比，建筑和城市设计被赋予了新的、更广泛的机会来营造空间形式多样化。现在，分散的、去中心化的、非集中化的城市形态远比过去更具有技术上的可行性。”[7]

在这个意义上，雅各布斯的思想虽然一开始是为了阻挡推土机的前进，在草根群体集会上发出的呼吁，但却被证明是领先于时代的。

---

1. 简·雅各布斯：《市中心是人民的》，载《财富》，1958年，登录日期2015年9月6日，http//fortune.com/2011/09/18/downtown-is-for-people-fortune-classic-1958/。
2. 简·雅各布斯：《城市经济学》，纽约：兰登书屋，1969年，第88—89页。
3. 简·雅各布斯：《分离主义问题：魁北克与主权斗争》，纽约：兰登书屋，1980年。
4. 简·雅各布斯：《城市与国家财富》，纽约：温特吉出版社，1984年；简·雅各布斯：《生存系统》，纽约：兰登书屋，1992年。
5. 简·雅各布斯：《经济的本质》，纽约：兰登书屋，2000年；简·雅各布斯：《即将来临的黑暗时代》，纽约：兰登书屋，2004年。
6. 安东尼·弗林特：《与摩西摔跤：简·雅各布斯如何挑战纽约营造大师并改变了美国的城市》，纽约：兰登书屋，2009年，第185页。
7. 大卫·哈维：《后现代主义的状况》，麻州剑桥：布莱克威尔，1990年，第75—76页。

第三部分：学术影响

9 最初反响

要点

- 雅各布斯的批评者嘲笑她提出的是一些不切实际的城市重建方案，并忽视了社会阶层等因素。
- 雅各布斯没有直接回应对她思想的批评，但对许多批评者的傲慢姿态非常反感。
- 城市重建在许多城市的毁灭性影响导致了一个共识，即《美国大城市的死与生》是一本恰逢其时的好书。

批评

《美国大城市的死与生》出版后，既收获了赞扬，也受到了批评。书籍出版几个月后，著名城市研究学者劳埃德·罗德文*在《纽约时报》撰文提出了一种两可的观点。“读者会强烈同意或反对这些观点，”罗德文写到，“但几乎没有哪个在读完书之后，不在看待自己的街道和社区时角度略有不同、稍微更加敏感。”[1]事实上，我们不可能忽视《死与生》的原创力和它影响的意义。雅各布斯曾效力过的《建筑论坛》的编辑们热情更加高涨：“任何领域里，一旦一些长期被接受的观念受到挑战，尤其是当这种挑战是以高智商为保障和以人文主义为基础的时候，不都很奇妙吗？”[2]无论你是爱它还是恨它，《死与生》都毋庸置疑地表明简·雅各布斯向城市规划的主导理论发起了一场重大挑战。

《死与生》的一个重要批评家是刘易斯·芒福德，他是城市历史学家，为《纽约客》杂志一个定期的建筑专栏撰文。芒福德也反对罗伯特·摩西的一些城市重建项目，并在听了雅各布斯在哈佛大学设计学院研究生院的演讲、读了她在《财富》杂志上的文章《市中心是人民的》之后，鼓励过雅各布斯。然而雅各布斯在她的书中批评一些顶尖城市规划者的同时也批评了芒福德的《城市文化》一书。于是芒福德针锋相对，写了一篇口气傲慢的文章《雅各布斯妈妈的居家良方》批评《死与生》。[3] 他贬斥雅各布斯是一个学术票友，一个业余爱好者："对雅各布斯夫人缺乏历史知识和学术敬畏心我毋庸赘言了，可是她对显而易见的事实的漠视也太频繁了。"[4]

一篇更加公平、更加礼貌的评论出自社会学家赫伯特·甘斯 * 发表在《评论家》杂志的文章。甘斯称赞《死与生》是"代表传统城市进行深入思考和充分想象的小册子。"[5] 但他也对他所谓的雅各布斯的"物理谬误"提出质疑，这个谬误促使雅各布斯"忽略了那些或带来活力或带来停滞的社会、文化和经济因素。"[6] 简言之，雅各布斯的分析没有考虑到其他变量，如社会阶层。他接着还说："在倡导对城市进行规划以刺激丰富的街道生活方面，雅各布斯不仅高估了规划在塑造行为方面的作用，而且她实质上还要求中产阶级采纳工人阶级的家庭生活、养儿育女和社会交往的方式。"[7] 甘斯认为这一视角限制了她的分析视野。

> "《死与生》从未登上过畅销书榜，但它一问世就触碰到了人们的神经。人们在所有地方谈论它，报纸社论、书评、教室、会议室、公开研讨会。雅各布斯的案例陈述非常有力，你不可能忽视《死与生》，即使你讨厌它也无济于事，而正如可以预见的那样，许多人确实如此，尤其是政治家、开发商，以及其他任何利益遭受雅各布斯攻击的人。"
>
> ——爱丽丝·斯帕伯格·阿莱克休：《简·雅各布斯：城市梦想家》

回应

对绝大多数批评她的人，雅各布斯都没有直接回应；她继续参与反对城市重建的行动，这强有力地表明她愿意为自己的论点而战。雅各布斯在她的第二本书《城市经济学》中进一步发展了许多最初在《死与生》中有所铺陈的观点。

《死与生》出版 40 多年后，雅各布斯在 2004 年的一次纽约城市学院的演讲中讲述了她对芒福德批评的感受。她嘲笑了芒福德自命不凡的性别歧视（其他批评家也抱有此种观点）："我认为他对这本书的反应不够理性……也许如果他生活在不同的年代，他就会明白女人未必都渴望得到庇护。他相信妇女是人类的一种女性附属品。"[8]

尽管如此，雅各布斯似乎受到了那些思想更加深邃、不带政治动机的批评家们的伤害。她的传记作者指出"她常常不能善待批评。"[9]尤其是她对赫伯特·甘斯在《评论家》上的评论"非常生气"，该篇批评在赞扬和批评中走了中立路线。[10]雅各布斯之前在波士顿时曾和甘斯见过面并交流过思想，1962 年甘斯在他自己的书《城中村》中集中讨论了波士顿北端社区居民经历的、与城市重建的斗争。[11]

冲突与共识

雅各布斯批评者们的普遍共识是她正确地指出了城市重建的负面影响。的确，它的毁灭性影响在全国各地任何一座城市都是显而易见的。然而，许多人也认为她的解决方案和建议会误导大家，甚或是完全错误的。一句话，许多人提出"是雅各布斯的分析使她的书精彩绝伦，而不是她的解决方案。”[12] 正如甘斯在他关于《死与生》的评论中所指出的那样，大多数中产阶级人士不一定渴望多样性，他们更喜欢郊区社区，而不是城市社区，他们也不热衷步行或使用公共交通。[13]

随着时间的推移，关于《死与生》的争论焦点发生了转变。当美国政治转向右翼时，一些左翼批评家认为保守派可能会欣赏她的观点。1982 年，政治学家马歇尔·伯曼写道：“这里与之相关且令人不安的是，新右翼的理论家多次引用雅各布斯，如同她是他们的守护神……在我看来，在她现代主义文本之下的是一种反现代主义的潜文本，是对自我可以牢牢扎根其中的家庭和邻里的一种怀旧情绪的暗流。”[14]

20 世纪 80 年代，《死与生》的怀旧情绪开始和保守主义的价值观、多愁善感交织在一起。这未必是雅各布斯的意图，但不断变化的社会环境产生了一些明显的共同之处——这有点不同寻常，因为雅各布斯时代的保守派几乎不可能会认为自己是城市的捍卫者。

---

1. 劳埃德·罗德温：“邻居是必需的”，《纽约时报》，1961 年 11 月 5 日。
2. 引自爱丽丝·斯帕伯格·阿莱克休：《简·雅各布斯：城市梦想家》，新泽西州新不伦瑞克：罗格斯大学出版社，2006 年，第 84 页。
3. 刘易斯·芒福德：“天际线：雅各布斯妈妈的居家良方”，《纽约客》，1962 年 12 月 1 日。
4. 芒福德：“天际线”。
5. 赫伯特·甘斯：“城市规划和城市现实”，《评论家》第 33 卷，1962 年 2 月，第

170—175 页。

6. 甘斯："城市规划和城市现实"。
7. 甘斯："城市规划和城市现实"。
8. 引自阿莱克休：《简·雅各布斯》，第 94 页。
9. 阿莱克休：《简·雅各布斯》，第 94 页。
10. 阿莱克休：《简·雅各布斯》，第 94 页。
11. 赫伯特·甘斯：《城市村民：意裔美国人生活中的群体和阶级》，纽约：格伦科自由出版社，1962 年。
12. 阿莱克休：《简·雅各布斯》，第 83 页。
13. 甘斯："城市规划和城市现实"。
14. 马歇尔·伯曼：《一切坚固的东西都烟消云散了：现代性经验》，纽约：企鹅图书，1982 年，第 320 页。

10 后续争议

要点

- 简·雅各布斯的思想给城市中产阶级化（一个社区成功地再生致使其原来的居民再也买不起它的过程）研究提供了新的关联，并影响了城市规划的其他形式。
- 新城市主义学派在一定程度上受到了《美国大城市的死与生》的启发。
- 雅各布斯的思想对城市研究及许多城市研究的一流学者，如理查德·塞内特*和莎伦·佐金*来说，都是变革性的。

应用与问题

在《美国大城市的死与生》中，简·雅各布斯抨击了自第二次世界大战结束以来占主导地位的城市规划思想。到1961年这本书出版的时候，诸如罗伯特·摩西这样的规划者手里的城市重建项目将要结束。城市面临着一个变革的时代，雅各布斯的观点即表明并催化了这些变化。城市中产阶级化的最初迹象已开始显现，到20世纪80年代许多城市社区的面貌发生了改变。回头来看，雅各布斯的思想作为这些变化的先兆重新获得了重视："城市中产阶级化是她作品的中心张力，但在1961年，雅各布斯还没有足够的词汇来说明它。现实中，她的哈德逊街并不是一个古色古香的村庄，而是一个充满活力的中间地带，脆弱地存在于不断扩张的后工业景观和日渐衰落的工业化景观之间。"[1]

处于城市中产阶级化最前沿的是年轻的中产阶级专业人士，他们要寻找在郊区无法找到的真实感和生活方式。他们把这种需求和雅各布斯在《死与生》中描述的族裔社区、本地企业、老旧的建筑物和适合步行的街道联系在一起。正如城市社会学家和城市中产阶级化的重要理论家莎伦·佐金所说："简·雅各布斯比任何人都更好地表达了这种新型的城市真实感的吸引力。"[2]《死与生》已经生发出新的意义，因为城市理论家们寻求阐释城市中产阶级化——以及白领们对刺激它发展的真实感的追逐。

> "如果类似简·雅各布斯这样的作家试图从现代资本主义同化中拯救空间，他们也会纪念城市演变的某个瞬间的、位于大都市的某个特定地点：在城市中产阶级化初期，一个19世纪工业的中心位置的城市景观，处在战后现代中心商业区或大学校园的边缘。"
>
> ——苏莱曼·奥斯曼：《褐色砂石建筑布鲁克林的发明》

思想流派

雅各布斯成为设计与规划运动兴起中最有影响力的人物之一，这个运动被称作新城市主义。这个松散定义的群体包括建筑师、规划师，和主张规划多样性、强调街头生活重要性并为实现这些目标提供具体建筑解决方案的学者们。[3]他们的核心原则记载于1993年的一份宪章："新城市主义代表大会认为，城市中心区投资的减少，不受地点限制扩张的蔓延，种族和收入导致的差距的日益扩大，环境的恶化，农业用地和荒野的丧失，以及社会中建筑遗产的侵蚀构成了和街区建设密切相关的挑战。"[4]

新城市主义建立在雅各布斯思想基础之上，并将其应用于当代的社会不平等和生态危机问题。

20 世纪 90 年代以来，新城市主义吸引了众多实践者，已经成为城市规划中最具影响力的运动之一。它的追随者们也没有少受批评。例如，城市理论家大卫·哈维 * 是一位研究资本主义与社会空间之间关系的学者，这种研究途径是那些利用马克思主义 * 社会经济学理论作为理论工具的研究中的典型。他就指责新城市主义者主要是满足年轻中产阶层专业人士的需要，而忽视了城市内部下层社会的困境。[5] 另一些人批评新城市主义的信徒，因为他们把他们的想法应用于小型城市、规划好的街区和没有并入大都会的社区；还有一些人则认为他们的项目感觉上是人为的、有太过严格的规划色彩。[6]

雅各布斯本人在 2001 年的一次采访中被问及新城市主义时似乎也呼应了这些批评："新城市主义者希望在他们开发的地方有活跃的中心，在那里人们会在出门办事及类似事情时互相碰见。然而，就我所见的他们的规划和所建的地方来看，他们似乎没有找准剖析这些心脏、这些中心的感觉。他们把它们当购物中心一样布置。彼此并不相连。"[7] 因此，尽管雅各布斯启发了新城市主义者们，但她却认为后者误解了自己的思想。

当代研究

雅各布斯的《死与生》影响了许多当代城市设计和规划领域的学者与实践者。她影响了美国两位重要的城市学家理查德·塞内特和莎伦·佐金，尽管他们也和她的一些思想保持着较大距离。塞内特对人们如何通过身体感官体验城市感兴趣。他使用类似于雅各布斯的方法研究格林尼治村的日常生活和社会交往。"和许多其他人一样，"塞内特写道，"早在我到达那里的20年之前，我就通过阅读简·雅各布斯的《美国大城市的死与生》了解了格林尼治村。"[8] 虽然塞内特对于雅各布斯的许多结论很不赞同，但是他在学术上深受雅各布斯的影响是显而易见的。

莎伦·佐金是另一位受雅各布斯影响的学者，尽管她也与她的思想有所分歧。在对城市中产阶级化的研究中，佐金进一步发展了雅各布斯关于城市与文化交互影响的一些思想。她还研究了被雅各布斯忽视的问题，特别是城市中的种族、阶级和社会不平等问题。然而，尽管佐金将自己与雅各布斯区分开来，但她认识到，《死与生》对于城市中产阶级化研究仍有持续的重要意义，而事实上，它已经预见到了这一变化："她把小而旧的建筑、低廉的租金与社区街道生活、专门化的低价商店，以及新的、有趣的经济活动联系在一起，换句话说，就是市中心的社会价值观。"[9]

---

1. 苏莱曼·奥斯曼：《褐砂石建筑布鲁克林的发明：城市中产阶级化与战后纽约真实性的找寻》，纽约：牛津大学出版社，2011年，第177页。
2. 莎伦·佐金：《裸城：真实城市空间的死与生》，纽约：牛津大学出版社，2010年，第16页。
3. 彼得·卡茨：《新城市主义：走向社区建筑》，纽约：麦格劳-希尔，1994年。
4. 新城市主义代表大会：《新城市主义宪章》，《城市读本》，第5版，理查德·T. 勒盖茨和弗雷德里克·斯托特编，纽约：劳特利奇，2011年，第357页。
5. 大卫·哈维：《新城市主义与社群主义陷阱》，载《哈佛设计杂志》第1卷，1997年，第68—69页。

6. 见托德·W. 布雷西编:《海边辩论:对新城市主义的批判》,纽约:里佐利,2002年。
7. 简·雅各布斯和比尔·施泰格瓦尔德:《城市景观:城市研究传奇简·雅各布斯论城市中产阶级化、新城市主义和她的遗产》,《理性》第33卷,2001年6月第2期。
8. 理查德·塞内特:《肉与石:西方文明的身体与城市》,纽约:W.W. 诺顿,1994年,第355页。
9. 莎伦·佐金:《权力的风景:从底特律到迪斯尼世界》,伯克利和洛杉矶:加利福尼亚大学出版社,1991年,第191页。

11 当代印迹

要点

- 《美国大城市的死与生》被广泛认为是城市研究和城市规划的经典著作。
- 《死与生》继续挑战着城市规划者，使他们建设能提高居民生活质量的城市。
- 关于雅各布斯思想的合法继承人，理论家和规划者之间的争论还在继续。

地位

简·雅各布斯的《美国大城市的死与生》毋庸置疑是一部城市研究经典。普遍的共识是，这本书在理论和实践两方面改变了理解城市的方法。《死与生》开启了战后美国和国际上关于城市功能的新理论争论，标志着城市的设计和规划方式发生了重大变化。雅各布斯的书会频繁出现在一长列各种学科的课程阅读清单上，并经常有章节被节选收入城市著作选集中，[1] 而且它几乎始终是城市研究教科书的讨论对象。[2]

在《死与生》出版后的数十年里，批评已经从有关城市重建的制度问题转向雅各布斯疏忽了的问题，如种族、阶级和社会不平等。社会学家莎伦·佐金和马克思主义地理学家大卫·哈维站在这不断变化的焦点的最前沿。雅各布斯最初因缺乏专业学历，缺少系统性的方法论和过分简化的文献而受到批评。[3] 但这些反对意见现在已经不重要了，而她对种族和阶级的漠视则造成了更多的问题。

例如，她对格林尼治村的理想化缺乏对不平等的分析："这就是为什么她的邻里风景看似田园牧歌的原因：这是一个黑人到达之前的城市。她的世界是从底层的稳固的工人阶级白人，一直到顶层的专业性的中产阶级白人。"[4] 所以，虽然《死与生》仍然被认为是一部经典，但对种族和阶级感兴趣的学者却只得另寻他途。

> "来自斯克兰顿的女孩站出来反抗摩西，挑战现状。现在，实质上几乎所有从事城市建设的人都追随她的规则。她的胜利铭刻在开发商、城市官员、游说团体和草根组织遵循的各种协议中，而一本本《美国大城市的死与生》摆上了全国各地市政厅的规划办公室的书架。"
>
> ——安东尼·弗林特：《与摩西摔跤：简·雅各布斯如何挑战纽约营造大师并改变了美国的城市》

互动

雅各布斯写作《死与生》的特定历史背景是当时城市重建正在蹂躏城市，一连串的错误假设影响了规划者的做法。雅各布斯向正统理论 * 发起了挑战，促使它失去了无可置疑的权威性。从此，《死与生》的许多核心理念成为城市设计与规划的新常识。随着城市研究内部的争论点早就更新换代，《死与生》对城市规划的批判已经成了正典。

有一群理论家和规划者继承和发扬了雅各布斯的思想，其中包括理查德·塞内特 * 和伦敦政治经济学院的城市研究教授理查德·伯德特 *，后者还是"开放城市"——一个基于民主和多样性原则的城市——的支持者。塞内特对雅各布斯思想的运用遵循了《死与生》的精神和意图；他援引雅各布斯来解释开放城市的理念。

他声称，雅各布斯“相信在一个开放城市里，就像在自然界中一样，社会和视觉形式会根据随机变化而发生变异；人们能够最好地吸收、参与和适应他所说的“城市时代”的变化。”[5]

新城市主义是由另一群受雅各布斯影响的理论家和规划者组成的。在他们规划的街区中，新城市主义者采纳了雅各布斯关于混合用途地区、短街区和行人友好型社区的许多建议。这些核心原则在其创始章程中概括了出来：“居民区在用途和人口方面应该是多样化的；社区应该既为行人与通行设计，也为小汽车设计；城市和小镇应由物理上轮廓清晰的、普遍可通达的公共空间和社区机构构成；城市空间应以颂扬本地历史、气候、生态和建筑实践的建筑和景观设计为框架。”[6]

尽管雅各布斯划清了自己和新城市主义的界限，[7]但《死与生》显然是新城市主义思想的基础文献。

持续争议

雅各布斯已经成为城市研究中一位受人尊敬的人物。然而，一边是众多的理论家和规划者声称受到她的影响，一边仍有一些争议存在。哈佛大学经济学家爱德华·格莱泽*是对雅各布斯及其准则提出挑战的当代学者之一。在他的《城市的胜利》一书中，格莱泽宣称：“这本书中的许多观点都借鉴了简·雅各布斯的才智，她知道你需要走在城市的街道上才看得清它的灵魂。”[8]尽管如此，格莱泽继而还是从理论和方法论上反驳了雅各布斯的一些结论。他坚称，雅各布斯“也会犯错，她太依赖于她的平地视角，而不去运用有助于从一个完整系统角度思考的概念性的工具。”[9]具体来说，格莱泽质疑了她的结论——即限制建筑物的高度和保护老旧居民区将使城市更加经济实惠——并且建议规划者应重新考虑高层公寓和摩天大楼的好处。

像开放城市鼓吹者和新城市主义者一样，格莱泽关注城市生活对环境的影响，但对城市的社会不平等问题不那么担心。对雅各布斯的各种不同反应一部分是因为大家将她的理念付诸实践的方式不同。格莱泽是一位城市经济学家，他以学者和理论家的身份审视城市；相比之下，新城市主义者大多是实践者，在规划的社区和建筑中具体体现雅各布斯的理念。格莱泽声称遵循雅各布斯的传统开展工作，开放城市的倡导者和实行新城市主义的规划者也这么说——但是可以肯定的是，每一名门徒得出的结论却是大相径庭的。

---

1. 理查德·T. 勒盖茨、弗雷德里克·斯托特编：《城市读本》，第 5 版，纽约：劳特利奇，2011 年；珍妮特·林、克里斯托弗·梅莱编：《城市社会学读本》第 2 版，纽约：劳特利奇，2012 年。
2. 马克·戈迪纳、雷·哈钦森和迈克尔·T. 瑞安：《新城市社会学》，第 5 版，科罗拉多州博尔德：韦斯特维尔出版社，2015 年；约翰·J. 麦休尼斯、文森特·M. 帕里略：《城市和城市生活》第 6 版，马萨诸塞州波士顿：培生，2013 年。
3. 刘易斯·芒福德：《天际线：雅各布斯妈妈的居家良方》，《纽约客》，1962 年 12 月 1 日。
4. 马歇尔·伯曼：《一切坚固的东西都烟消云散了：现代性经验》，纽约：企鹅图书，1982 年，第 324 页。
5. 理查德·塞内特：《开放城市》，《城市时代》，2006 年，第 2—3 页。
6. 新城市主义代表大会：《新城市主义宪章》，见《城市读本》，第 5 版，理查德·T. 勒盖茨和弗雷德里克·斯托特编，纽约：劳特利奇，2011 年，第 357 页。

7. 简·雅各布斯和比尔·施泰格瓦尔德:《城市景观:城市研究传奇简·雅各布斯论城市中产阶级化、新城市主义和她的遗产》,《理性》第 33 卷,2001 年第 2 期,第 48—55 页。
8. 爱德华·格莱泽:《城市的胜利:我们最伟大的发明如何让我们更富有、智慧、绿色、健康和幸福》,纽约:企鹅图书,2011 年,第 11 页。
9. 格莱泽:《城市的胜利》,第 11 页。

12 未来展望

要点

- 随着世界各地城市人口持续增加,《美国大城市的死与生》的影响可能会继续增大。
- 《死与生》祛除了二战后一个时期典型的城市规划思想和实践，且将继续承当城市规划理论基础文献的角色。
- 《死与生》是一部奠基性的作品，它批判了城市重建，描述了城市中的社会交往，预示了城市中产阶级化。

潜力

简·雅各布斯的《美国大城市的死与生》可能会对美国以外的理论家和规划者产生影响。截至 2015 年，世界上的大多数人口都生活在城市地区——这是全球历史上的第一次。[1] 世界各地的城市在快速发展，速度比以往任何时候都更快，而且没有停止的迹象。联合国预估如果以目前的速度继续增长，世界城市人口将每隔 38 年翻一番。[2] 显然，这些情况要求人们更清楚地了解是什么使城市顺利运转，以及该如何规划城市，而这正是雅各布斯著作的意义之所在。

全球的城市都因吸纳简·雅各布斯关于经济学与空间的关系的洞见而受益。当代城市理论家，例如理查德·弗罗里达*，都强调城市对经济创新的重要性。[3] 弗罗里达和其他许多人，包括那些学术界之外的人，都认识到城市空间在培养全球经济竞争所必需的创造力和多样性方面发挥着关键作用。马尔科姆·格拉德威尔*是好几本社会科学著作的作者，他在2000年写道："今天重读《死与生》，一定会被中间这些年赋予她的论点一种新的意想不到的重要性所震惊。毕竟，谁有直接的兴趣去创造多样化的、至关重要的空间来培养创造力和意外收获呢？雇主们有。在《死与生》出版迎来40周年纪念之际，它再度流行，成为工作空间设计方面的启蒙读本。"[4]

> "但在一些地方，比如阿根廷，一部简·雅各布斯作品的西班牙语译本过去只能藏身于图书馆的书库中，而现在她正在不断地收获追随者，就像一个方兴未艾的保护运动在富有不破不立精神的布宜诺斯艾利斯寻求一个哲学的根基……在所有这些地方，一切都在提醒我们，简·雅各布斯在《死与生》出版50年后仍然是一个重要的人物，其影响力在不断发酵。"
>
> ——麦克斯·佩奇和蒂莫西·门内尔：《重新审视简·雅各布斯》

未来方向

一群贯彻雅各布斯思想并应用她的方案的主要理论家和规划者自称是新城市主义者。[5] 总体而言，"新城市主义者认为《死与生》是他们的运动最重要的、始创意义的理论，他们不仅接受了雅各布斯关于城市密度、适合步行的街区和'街道眼睛'的概念，还接受了她所信奉的混合各种用途、建筑物和人流的理念。"[6] 新城市主义者在他们的规划和设计中实践了雅各布斯的许多建议，特别是培育社会交往和用途多样化的建议。

然而，在2001年的一次采访中，雅各布斯划出了与这场运动的距离。[7] 一些理论家和规划者批评新城市主义把雅各布斯的理论

和城市环境相剥离，并将其应用于更像郊区或小城镇的社区：[8]
“新城市主义在应用雅各布斯思想方面的主要问题在于，多样性能否作为规划和城市设计的理论先决条件被规划。”[9]

具有讽刺意味的是，尽管雅各布斯批评了她那个时代的城市规划，并主张采取更有机、更自发的方式对待城市，但她的思想现在却在更小的非城市社区的新型规划中找到了用武之地。当然，尽管它们在特质上可能不是城市，但是它们至少在尝试表现雅各布斯的思想。

小结

《死与生》从根本上改变了城市研究和城市规划，而且其影响力仍在继续。当它在1961年出版时，格林尼治村正在进行一场反对建筑规划师罗伯特·摩西的城市重建项目的居民区运动，雅各布斯本人也参与其中。这些事件塑造了《死与生》，而且该运动的成功证实了书中反对城市重建的效果。雅各布斯在揭穿摩西和其他规划者实施的正统派*理论的同时，提出一系列通过促进社会交往保障城市运行的替代性观点。自那以后，她对混合用途空间、适合步行居民区和历史保护的建议已经在城市规划中被广泛接受。

《死与生》的历史意义和广泛影响使其成为城市研究和规划中不可或缺的著作。从那以后，它已经成为一些当代理论家和规划者的基础性文献，如与新城市主义有关的规划者。即使有如爱德华·格莱泽这样的学者很有礼貌地不赞同雅各布斯的思想，《死与生》仍然是这一领域一块重要的试金石。[10] 她对格林尼治村的分析还预见到了城市中产阶级化的出现，这一现象在《死与生》出版后的数十年间改变了大量的城市居民区。包括社会学家莎伦·佐金在内的学者们认为，雅各布斯忽视了种族和阶级的不平等，这些不平等现象随着城市中产阶级化而涌现出来，但正是在这里，再次说明雅各布斯联结城市和文化的能力是具有开创意义的。[11]

政治学家马歇尔·伯曼写到，雅各布斯在《死与生》中描述了“街道现代主义”。[12]他的意思是，她捕捉到了都市环境中人的混合所产生的兴奋、创造力和活力。她的观点与摩西这样的规划者恰恰相反，摩西他们试图摧毁城市，因为他们害怕街道和密集的人群。但是这些恐惧被证明是没有根据的，而随着雅各布斯的观点得到人们的青睐，摩西和他的同类所支持的项目在正式开始之前就已经注定了失败的命运。在雅各布斯与摩西在格林尼治村对决中争论不休的 50 多年后，城市依然存在。随着城市继续增长，雅各布斯的影响力看起来也将不断增长。

---

1. 马克·戈迪纳、雷·哈钦森和迈克尔·T. 瑞安：《新城市社会学》，第 5 版，科罗拉多州博尔德：韦斯特维尔出版社，2015 年，第 9 页。
2. 戈迪纳、哈钦森和瑞安：《新城市社会学》，第 10 页。
3. 理查德·弗罗里达：《创新阶级的兴起》，纽约：基本图书，2002 年。
4. 马尔科姆·格拉德威尔：《工作设计》，《纽约客》，2000 年 12 月 11 日。
5. 彼得·卡茨：《新城市主义：迈向社区建筑》，纽约：麦格劳–希尔，1994 年。
6. 马提亚斯·温特：“简·雅各布斯著作《美国大城市的死与生》对城市规划职业的重要性”，《公共事务新视野》第 1 卷，2009 年春。
7. 简·雅各布斯和比尔·施泰格瓦尔德：“城市景观：城市研究传奇简·雅各布斯论城市中产阶级化、新城市主义和她的遗产”，《理性》第 33 卷，2001 年第 2 期，第 48—55 页。

8. 见麦克斯·佩奇和蒂莫西·门内尔编：《重新审视简·雅各布斯》，芝加哥：美国规划协会，2011 年。
9. 温特："《美国大城市的死与生》的重要性"。
10. 爱德华·格莱泽：《城市的胜利：我们最伟大的发明如何让我们更富有、智慧、绿色、健康和幸福》，纽约：企鹅图书，2011 年。
11. 莎伦·佐金：《裸城：真实城市空间的死与生》，纽约：牛津大学出版社，2010 年。
12. 马歇尔·伯曼：《一切坚固的东西都烟消云散了：现代性经验》，纽约：企鹅图书，1982 年，第 314 页。

术语表

1.《建筑论坛》：1892 年至 1974 年美国出版的建筑与家居设计杂志。

2. **芝加哥学派**：芝加哥大学社会学系全体教师组成的学派，最著称于世之处是他们的城市社会学研究。20 世纪 20 至 30 年代，他们在城市生活方面开展了富有开创性的研究。芝加哥社会学家进行了许多关于城市亚文化和生活方式的人种志研究，而他们的理论是将城市的社会生活比作一种生态系统。

3. **横穿布朗克斯高速公路**：一条主要的汽车公路，由罗伯特·摩西设计，建于 1948 至 1972 年间。它的修建涉及拆除布朗克斯南部成千上万的房屋和企业，对其居民区造成了长期的毁灭性的后果。

4. **分散主义者**：19 世纪以来的一群城市理论家和规划者，他们寻求通过分散人口和已建成的城市环境来解决城市生活中的社会和环境弊病。

5. **花园城市**：19 世纪末由埃比尼泽·霍华德提出的一种乌托邦式的城市规划理念，作为当时人口稠密城市的另一种选择。据说，它在设计上将城市和乡村生活的最佳特征相互结合。

6. **城市中产阶级化**：白领（中产阶级）专业人士向之前恶化的城市居民区迁移的现象。这会增加房地产价值和居住成本，迫使以前的低收入居民离开，并引入白领人士在饮食、艺术、音乐和娱乐方面的文化品位。

7. **格林尼治村**：曼哈顿下城的一个社区，为作家、艺术家、音乐家和持不同政见者建立了一个享有美誉的波希米亚式、思想自由的避风港。

8. **历史保护运动**：在城市地区保护、保存和修复具有历史意义的建筑物、纪念碑和遗迹的运动。这一运动受到了社区反对城市重建拆除工程的推动。

9. **反对下曼哈顿高速公路联合委员会**：一场成功阻止建设一条由罗伯

特·摩西设计的汽车高速公路项目的地方运动，该高速公路计划穿过曼哈顿的居民区和商业区。

10. **下曼哈顿高速公路**：由规划师罗伯特·摩西提出的建造一条汽车高速公路的计划，该高速公路将连接曼哈顿下城东西两端。项目将需要拆除今天的 SoHo 和小意大利社区。社区的反对迫使该项目于 1962 年取消。

11. **马克思主义**：既是一种社会学分析方法论，也是一种历史发展理论。受德国经济学家和政治哲学家卡尔·马克思著作的启发，它的传统重点是阶级冲突、行为的经济决定论，以及对资本主义（以利润为导向）经济进行系统而深刻的批判。

12. **混合用途开发区**：将工业、商业、住宅和文化的空间混合在一起的地区或地带。这种组合的优点包括缩短目的地间的距离，更强烈的社区特征，以及拥有更多对自行车和行人友好的街道。混合用途地区越来越紧密地和城市中产阶级化相联合。

13. **新城市主义**：20 世纪八九十年代兴起的城市设计和规划运动。新城市主义者设计了许多将适于步行的街道、市中心区、历史保护和环境可持续性有机结合的社区。

14. **开放城市**：伦敦政治经济学院的城市学者理查德·伯德特和理查德·塞内特提出的基于民主和多样性原则的理想城市。它的核心特征是通过式的地域、不完整的形式、发展的可叙述性和民主的空间。

15. **正统派**：与传统的或普遍接受的规则或惯例相一致的事物。

16. **正统理论**：普遍接受的理论、信条或惯例。

17. **后现代建筑**：肇始于 20 世纪 70 年代后期的运动，参与的建筑师们试图脱离主导战后建筑的国际主义风格。与现代主义建筑相反，后现代主义建筑融合了自我指涉的巧智和非功能的装饰。

18. **辐射城市**：建筑师勒·柯布西耶的理想城市模型，发展于 20 世纪二三十年代。这个理想城市由高层建筑、高效的交通流和富足的绿色空间组成。

19. **第二波女权主义**：20 世纪 60 年代末开始的妇女社会运动，延续了早期妇女运动中争取选举权、非歧视、生育权和社会平等权的斗争。

20. **社会学**：研究人类社会的历史、形成和结构的科学。

21. **斯帕迪纳高速公路**：一条拟建的穿过多伦多市中心区的高速公路。由于公众的反对，该项目于 1971 年被取消。

22. **《纽约客》**：1925 年开始出版的一份周刊，包括小说、新闻报道、诗歌、批评、讽刺和漫画等栏目。

23. **城市规划**：致力于城市基础设施、交通、通信和公共福利的城市设计技术流程。

24. **城市重建**：主要指第二次世界大战后经常在城市中实施的重建项目。城市重建一般都包括拆毁城内的居民区来修建汽车高速公路。

25. **城市社会学**：研究城市环境中的社会机制的科学。

26. **城市研究**：一种结合多种学科研究城市及其郊区的学术领域。最常见的次级课题包括城市经济学、城市规划、城市政治、城市交通和城市社会学。

27. **华盛顿广场公园**：纽约市格林尼治村社区中心地带的公共广场和聚会地。自 19 世纪下半叶以来，这个公园一直是音乐家、艺术家、诗人和艺人们聚集与表演的重要地点。

28. **华盛顿广场公园委员会**：由谢莉·海耶斯于 1952 年创立的团体，旨在阻止纽约市将机动车通行延伸并穿过格林尼治村华盛顿广场公园的计划。

29. **第二次世界大战**：1939 年至 1945 年间的全球性冲突，涉及世界诸多大国和遍及全球的许多其他国家。

人名表

1. **简·亚当斯**（1860—1935），社会改革家，芝加哥赫尔馆社会改革中心的联合创始人。她的社会工作提高了人们对城市贫困和公共卫生问题的认识。

2. **马歇尔·伯曼**（1940—2013），纽约城市学院和纽约城市大学研究生中心政治学杰出教授。他最著名的成就是从马克思人文主义视角分析现代性。

3. **理查德·伯德特**，伦敦政治经济学院的城市研究教授。他在发展“开放城市”理念方面发挥了核心作用。

4. **欧内斯特·伯吉斯**（1886—1966），芝加哥大学城市社会学家。他最广为人知的成就是城市同心圆发展理论。

5. **蕾切尔·卡森**（1907—1964），海洋生物学家和自然资源保护者。她的《寂静的春天》一书揭露了杀虫剂滥用带来的不良影响，并促使了环保运动的兴起。

6. **理查德·弗罗里达**（1957 年生），城市理论家和多伦多大学罗特曼管理学院教授。他最著名的论述是城市如何才能刺激“创意阶层”的经济创新。

7. **贝蒂·弗里丹**（1921—2006），女性主义作家，全国妇女组织的联合创立人、第一任主席。她的著作《女性的奥秘》经常被誉为是 20 世纪 60 年代第二波女性主义浪潮的催化剂。

8. **赫伯特·甘斯**（1927 年生），社会学家，1971 年至 2007 年任教于哥伦比亚大学。他的研究跨越了广阔的领域，包括城市重建的影响、穷人的生活和新媒体的运作机制等。

9. **帕特里克·格迪斯爵士**（1854—1932），苏格兰知识分子和城市规划先驱者。他阐述了地区规划的思想，反对在他那个时代居于统治地位的网格状城市规划。

10. **马尔科姆·格拉德威尔**（1963 年生），记者，自 1996 年起担任《纽约客》专职作家。他的 5 部书探讨了学术性社会科学的惊人影响力，全都登上了《纽约时报》的畅销书榜。

11. **爱德华·格莱泽**（1967 年生），哈佛大学经济学教授，他的研究探讨了城市如何推动经济繁荣和环境的可持续发展。

12. **迈克尔·哈林顿**（1928—1989），作家，美国民主社会主义党创始人。他的第一部书《另一个美国》帮助推动了 20 世纪 60 年代全美范围内的反贫困斗争。

13. **大卫·哈维**（1935 年生），纽约城市大学人类学和地理学杰出教授。他因为运用马克思主义原理分析资本主义和社会空间之间的关系而著称。

14. **谢莉·海耶斯**（1912—2002），纽约格林尼治村的社区活动组织者。她创立了华盛顿广场公园委员会，与罗伯特·摩西修筑穿越下曼哈顿高速公路的计划作斗争。

15. **埃比尼泽·霍华德**（1850—1928），英国城市规划理论家。他展望并设计了基于乌托邦思想的花园城市，将城市和乡村生活的最佳方面结合在一起。

16. **勒·柯布西耶**（1887—1965），瑞士裔法国建筑师和城市规划师。他的建筑理念影响了 20 世纪 20 至 30 年代的巴黎再设计过程，他设计的建筑散布于世界许多地方。

17. **罗伯特·摩西**（1888—1981），城市规划师和政府官员，1922 年至 1968 年间，他担任过纽约市的众多职位。他以“营造大师”而著称，在 20 世纪中叶改变了纽约及其周边郊区的面貌。

18. **刘易斯·芒福德**（1895—1990），研究城市的美国作家和《纽约客》杂志建筑评论家。他最著名的作品是《城市发展史》，该书获得了 1962 年的国家图书奖（非小说类）。

19. **拉尔夫·纳达尔**（1932 年生），消费者权益保护者、环境保护活动家和人道主义者。1982 年至 2008 年，他曾 5 次参选美国总统。

20. **罗伯特·帕克**（1864—1944），城市社会学家，1914至1933年任教于芝加哥大学。他的研究探讨了城市生态、社会解体、种族关系、移民和同化等诸多问题。

21. **劳埃德·罗德文**（1919—1999），麻省理工学院城市研究教授，麻省理工和哈佛城市研究联合中心联合创始人。他著有11部著作，并在20世纪50至60年代的城市规划中发挥了重要作用。

22. **萨斯基亚·萨森**（1947年生），哥伦比亚大学社会学教授。她一直是全球化和城市研究的先锋人物，她的书被翻译成了21种文字出版。

23. **理查德·塞内特**（1943年生），伦敦政治经济学院社会学荣休教授。他撰写了许多著作论述现代社会中城市、社会阶层、公共文化和工作的发展等问题。

24. **威廉·H. 怀特**（1917—1999），社会学家、城市学家和组织分析专家。最著名的是他1956年出版的著作《有组织的人》，已经销售了200多万册。

25. **路易斯·沃思**（1897—1952），芝加哥大学社会学家和芝加哥学派的领军人物。他最著名的研究是作为一种生活方式的城市主义。

26. **莎伦·佐金**，布鲁克林学院和纽约城市大学研究生中心社会学教授。她的著作研究了城市中产阶级化和城市与文化间的关系（主要是纽约）。

WAYS IN TO THE TEXT

KEY POINTS

- Jane Jacobs (1916–2006) was a US journalist who criticized urban renewal*—the policy of reconstructing cities, frequently by demolishing neighborhoods for the construction of automotive highways—and postwar city planning.
- *The Death and Life of Great of American Cities* exposed the failures of urban planning* (the process of designing cities with concerns for infrastructure, transportation, communications, and public welfare).
- Jacobs presented an alternative method for understanding cities based on the firsthand observation of social interaction. She proposed ways to make cities more diverse, walkable, and densely concentrated.

Who Was Jane Jacobs?

Jane Jacobs, the author of *The Death and Life of Great of American Cities* (1961), was born in 1916 in Scranton, Pennsylvania. She moved to New York City in 1934 to become a journalist, writing for the journal *Architectural Forum** and other magazines. In the late 1950s, Jacobs helped lead a movement to save Lower Manhattan—the southern part of New York City's largest island, with neighborhoods such as East Village and Chinatown, as well as the World Trade Center—from "urban renewal."[1] This eventually stopped plans that would have destroyed several neighborhoods in order to construct new roads. She published *The Death and Life of Great American Cities* in 1961, and its criticism of urban planning made an immediate impact by exposing the failures of urban renewal after World War II.* Many urban planners have since adopted Jacobs's ideas for making cities more diverse, walkable, and densely concentrated.

Jacobs moved to Toronto in 1968, where she joined a local movement opposed to the Spadina Expressway,* an urban renewal project that would have demolished numerous homes, parks, and small businesses.[2] Just as in New York City, this local movement succeeded in canceling the expressway's construction. Jacobs lived in Toronto for the rest of her life, and wrote six more books before her death in 2006, mainly about cities and economics. In the 1970s she became an advocate for the independence of Canada's French-speaking region of Quebec, publishing a book about the issue of Quebec separatism in 1980. However, *Death and Life* continues to be her most influential work.

What Does *Death and Life* Say?

Death and Life challenged dominant ideas of city planning and policy. Jacobs argues that urban planners destroy great cities because they do not consider how people live in them, and offers alternative ideas about how cities work developed by observing interactions in the streets. She insists that diversity, concentration, and mixture make cities great. In contrast, separation and standardization were the central principles of urban planning. Whereas planners assumed density and diversity created chaos, Jacobs sees these as sources of order and safety.

Ignoring the dynamics of city life, planners concocted proposals from urban theory. Jacobs sees these as fantasies that promised a better way of life but in reality accelerated deprivation and decline—without considering how millions of people actually inhabit and interact in urban spaces. Jacobs insists that these interactions characterize the life of the city. Urban planning could only succeed when its proposals took these social dynamics into account.

When she wrote *Death and Life*, Jacobs lived in in New York City's Greenwich Village,* a characterful neighborhood in Lower Manhattan, where her analysis was shaped by her observations of people and their social interactions. Jacobs also teamed up with other residents to save their neighborhoods from urban renewal projects. New York's "master builder" Robert Moses* planned to build an expressway through Manhattan. However, Jacobs and her neighbors rallied to stop Moses' plan. Their struggle was urgent because the expressway would have led to the demolition of numerous neighborhoods. Jacobs's fight against urban renewal was not just intellectual—it was political and personal. Moses' proposal would have impacted Jacobs because she lived in Greenwich Village, a neighborhood Moses condemned as a "slum," but that today is a hub for New York music, arts, and culture.

Death and Life offers several suggestions for urban planning alternatives. Jacobs outlined four conditions to create diversity in any city:

- First, a district should support a mixture of uses.While urban planners segregated commercial, residential, industrial, and cultural spaces, Jacobs maintained that city life improved when these different functions mixed.
- Second, blocks of streets must be as short as possible to

make them easier to walk and promote interaction.

- Third, districts should include a mix of new and aged buildings. Modernist urban planning assumed newer was always better, but older buildings maintain a sense of continuity on the streets.
- Finally, cities should foster a dense concentration of people. Urban planners held that large crowds were undesirable or even dangerous. Jacobs, on the other hand, believed density and mixture make cities safer and more enjoyable.

Death and Life has sold more than 250,000 copies and undergone six translations.[3] Jacobs writes in a pithy, accessible style that reflected her populist stance—and though she was not an academic, *Death and Life* forever changed the discipline of urban studies* (an academic field focusing on the economics, planning, politics, transportation, and sociology* of urban environments).

Why Does *Death and Life* Matter?

Jacobs's criticism of urban planning exposed its many failures and reshaped how people understand cities; it also generated new insights into the process. Many cities answered her call to mix residential, commercial, industrial, and cultural spaces. Though Jacobs was not by any stretch a planner, *Death and Life* marked a turning point in urban studies and planning.

Real-world conflicts over urban space informed the analysis of *Death and Life*. Jacobs and other Greenwich Village residents attacked Robert Moses' vision, even though he ranked as the most influential urban planner of the times. In defeating Moses' proposal for a Lower Manhattan Expressway,* their movement made history—and shaped the analysis of *Death and Life*.[4] While Moses' plans usually revolved around automobiles and traffic, Jacobs reminded him and his colleagues that people should come first. Moses saw no value in street life; Jacobs argues that streets made cities great. Moses demolished old neighborhoods he saw as "slums"; Jacobs insists that these held more value and function than new suburbs. *Death and Life* stimulated academic debate while changing urban policy and politics in the process.

Jacobs offers several recommendations that many cities have since adopted.[5] She was among the first to suggest that mixed-use development,* with districts blending industrial, commercial, residential, and cultural space, improved city life. This countered the orthodoxy* (that is, the generally accepted practice) that cordoned off zones with different functions. Jacobs argues that a mix of aged and new buildings made for better streets. This defense of old buildings gave rise to an urban movement for historical preservation*—a movement to protect, preserve, and restore buildings, monuments, and objects with historic significance in urban areas. Even though urban planners saw large groups of people as a danger, Jacobs insists on the importance of these dense concentrations, arguing that this mix of people provided all cities with a source of vitality and creativity. Jacobs also maintains that this made cities safer. Mixture and density create "eyes on the street" that watch over neighborhoods.[6] Instead of urban chaos, Jacobs sees it as "organized complexity," giving a sense of order.[7]

But urban planners, fearing congregation, separated and isolated people.

The ideas behind *Death and Life* have shaped urban policy, influenced historical preservation,[8] and highlighted the important economic functions that social interactions provide.[9] Where bureaucrats saw slums, decay, and expressways, Jacobs sees vitality, possibility, and a road to a different urban vision.

---

1 Anthony Flint, *Wrestling with Moses: How Jane Jacobs Took on New York's Master Builder and Transformed the American City* (New York: Random House, 2009), 3–28.

2 Flint, *Wrestling with Moses*, 182.

3 Stephen Ward, "Obituary: Jane Jacobs." *The Independent*, June 3, 2006, accessed August 29, 2015, http://www.independent.co.uk/news/obituaries/ jane-jacobs-6099183.html.

4 The links between *Death and Life* and Jacobs's activism in Greenwich Village are explored in Flint, *Wrestling with Moses*, 95–135.

5 For an assessment of Jacobs's legacy, see Roberta Brandes Gratz, *The Battle for Gotham: New York in the Shadow of Robert Moses and Jane Jacobs* (New York: Nation Books, 2010), 256–76.

6 Jane Jacobs, *The Death and Life of Great American Cities* (New York: Vintage Books, 1992), 35.

7 Jacobs, *Death and Life*, 429–39.

8 Gratz, *The Battle for Gotham*, 25–6.

9 Gratz, *The Battle for Gotham*, 266–8.

SECTION 1
INFLUENCES

MODULE 1

THE AUTHOR AND THE HISTORICAL CONTEXT

KEY POINTS

- *The Death and Life of Great American Cities* had an immediate impact with its searing condemnation of urban planning.* It continues to influence ideas and policies about city life.
- Jane Jacobs's observations of social interaction in Greenwich Village shaped her understanding of cities.
- While writing *Death and Life*, Jacobs joined Greenwich Village residents to save their neighborhoods from the architect and planner Robert Moses' reconstruction projects, which often proposed that neighborhoods be demolished to allow for the building of automobile infrastructure.

Why Read This Text?

Jane Jacobs's *The Death and Life of Great American Cities* (1961), with its blistering critique of architects such as Robert Moses—the "master builder" who transformed New York and its surrounding suburbs in the mid-twentieth century—challenged core ideas that dominated city planning and urban policy in the United States following World War II.* The work also instigated a paradigm shift (that is, a radical reappraisal) in urban planning after its 1961 publication. Jacobs developed an alternative view of city life, focusing on how people on the streets interact. Many cities have since implemented her proposals to mix residential, commercial, industrial, and cultural spaces.[1]

Death and Life presents a devastating assessment of urban planning in the period following World War II. It also anticipates many shifts in theory and policy that have occurred in the decades since the book's release. Jacobs's warnings about planning cities around automobile traffic have proven especially prophetic. By the time Jacobs wrote *Death and Life*, it had become clear that building more highways and expressways did not reduce traffic the way Moses and other planners promised. Jacobs saw that planning cities around cars and trucks devastated residential and downtown districts.[2] The erosion of communities and neighborhoods had become evident in the New York district of the South Bronx with the construction of the Cross Bronx Expressway.* Jacobs's criticism of Moses and other planners gave rise to an alternative urban transportation approach. Ideas like these from *Death and Life* have significantly influenced more contemporary forms of urban planning.[3]

> "The Death and Life of Great American Cities *hit the world of city planning like an earthquake when it was published in 1961.The book was a frontal attack on the planning establishment, especially on the massive urban renewal* projects that were being carried out by powerful redevelopment bureaucrats like Robert Moses* in New York. Jacobs derided urban renewal as a process that served only to create instant slums."*
>
> ——Richard T. Le Gates and Frederic Stout, *The City Reader*

Author's Life

Born in the city of Scranton in the US state of Pennsylvania in

1916, Jane Jacobs moved to New York City with her sister in 1934.[4] Although she was a professional journalist and writer, Jacobs never completed a degree or worked at a university.[5] She began writing about cities and architecture for *Architectural Forum** in the 1950s,[6] and when she wrote *Death and Life*, Jacobs lived in a renovated townhouse in the Lower Manhattan neighborhood of Greenwich Village.[7] *Death and Life* drew from her knowledge of cities, planning, and architecture, but also included what she observed from her home at 555 Hudson Street. In *Death and Life*, she wrote, "The stretch of Hudson Street where I live is each day the scene of an intricate sidewalk ballet."[8]

While writing *Death and Life*, Jacobs helped lead the movement to stop the construction of a Lower Manhattan Expressway.* The city's autocratic planner, Robert Moses, had conceived this highway, which would have bisected Washington Square Park,* a central place of recreation and cultural expression for village residents and those drawn to the neighborhood's bohemian character.[9] Jacobs chaired the Joint Committee to Stop the Lower Manhattan Expressway*—and was once arrested for destroying the stenographer's notes in protest at a public meeting in 1958.[10] A year after the publication of *Death and Life*, New York City officials were convinced by the movement to reject Moses' plan.[11]

Author's Background

After World War II, dominant ideas in urban planning involved slum clearance, the construction of high-rise public housing, and the creation of highways that linked cities with suburbs. This sort of urban planning separated residential, commercial, industrial, and cultural spaces. It razed whole neighborhoods for the sake of expressway traffic. And even though evidence of their failure mounted, these ideas had congealed into orthodoxy. The urban renewal movement developed over several decades from design and planning theories. Its roots took shape during the nineteenth century, a time when cities were seen as polluted, crowded, and generally undesirable places to live. Planners of that era introduced proposals and designs for utopian (that is, visionary and ideal) living and working spaces that sought to avoid the city's worst features.[12] They also presented urban renewal as a path to improve the economy and social life of cities, but Jacobs saw it as leading to the demise of everything that made cities work.

Contrary to the urban planners of her time, Jacobs argued that the life of cities stemmed from how people mixed on the streets. She contended that cities should include mixed-use spaces that combined industrial, commercial, residential, and cultural functions, while urban planners separated these into distinct districts. Her prescription for great cities also included short blocks that were easy to walk, a blend of old and new buildings, and a dense concentration of people. Where bureaucrats resorted to abstract theory, she relied on keen observation to inform her perspective: "People who are interested only in how a city 'ought' to look and uninterested in how it works will be disappointed by this book."[13] *Death and Life* mounted an unsparing attack on the ideas and policies that had held unquestioned acceptance among urban planners. It dashed their questionable projects and eventually eroded their false assumptions.

---

1 A contemporary assessment of Jacobs's legacy can be found in Roberta Brandes Gratz, *The Battle for Gotham: New York in the Shadow of Robert Moses and Jane Jacobs* (New York: Nation Books, 2010), 256–76.

2 Anthony Flint, *Wrestling with Moses: How Jane Jacobs Took on New York's Master Builder and Transformed the American City* (New York: Random House, 2009), 62–3.

3 Gratz, *The Battle for Gotham*, 274–6.

4 Flint, *Wrestling with Moses*, 3–4.

5 Flint, *Wrestling with Moses*, 8–9.

6 Flint, *Wrestling with Moses*, 18.

7 Flint, *Wrestling with Moses*, 99.

8 Jane Jacobs, *The Death and Life of Great American Cities* (New York: Vintage Books, 1992), 50.

9 Flint, *Wrestling with Moses*, 62.

10 Flint, *Wrestling with Moses*, xiv.

11 Flint, *Wrestling with Moses*, 158–9.

12 Flint, *Wrestling with Moses*, 20.

13 Jacobs, *Death and Life*, 14.

MODULE 2
ACADEMIC CONTEXT

KEY POINTS

- Jane Jacobs challenged two schools of thought in urban planning: that of the Decentrists* (a group of urban theorists and planners from the nineteenth century who sought to reform the social and environmental ills of city life by decentralizing the population and built environment of cities); and that of the disciples of the influential Swiss French architect Le Corbusier.*
- Although it expressed concerns similar to hers, Jacobs did not engage with the Chicago School,* best known for its urban sociology (the study of the social constitution of urban environments). Associated with the sociology department at the University of Chicago, the Chicago School produced several groundbreaking studies of urban life in the 1920s and 1930s.
- While most of the observations in *The Death and Life of Great American Cities* came from New York City, Jacobs also noted that the same problems plagued other American cities.

The Work in Its Context

In *The Death and Life of Great American Cities*, Jane Jacobs identified two schools of thought among urban planners. The first consisted of "Decentrists," and as the name implies, they sought to decentralize cities and disperse their people. Their main ideas came from the English urban theorist Ebenezer Howard* and the Scottish urban planner Sir Patrick Geddes,* who was noted for developing ideas that offered alternatives to the gridiron plans that dominated cities in his time. Conditions in London during the late nineteenth century horrified Howard, who envisioned a "Garden City"* where people would create self-sufficient small towns in the countryside. Jacobs described how Howard's hostility towards cities shaped his urban planning: "He not only hated the wrongs and mistakes of the city, he hated the city and thought it was an outright evil and an affront that so many people should get themselves into an agglomeration. His prescription for saving the people was to do the city in."[1]

Writing in the early twentieth century, Geddes wanted to extend Howard's garden-city ideal into entire regions, distributing such municipalities outside urban areas.

The second school of thought derived from the famous Swiss French architect Le Corbusier, whose ideas reshaped Paris in the 1920s and 1930s. His ideal Radiant City* called for a succession of skyscrapers standing in a park. As Jacobs wrote, "His city was like a wonderful mechanical toy... It was so orderly, so visible, so easy to understand. It said everything in a flash, like a good advertisement."[2] While the Decentrists influenced planning outside the city, Le Corbusier's ideas informed what went on inside it. His influence appears most evident in high-rise office buildings and low-income housing projects.

> *"I have been making unkind remarks about orthodox* city planning theory, and shall make more as occasion arises to do so. By now, these orthodox ideas are part of our folklore. They harm us because we take them for granted."*
>
> ——Jane Jacobs, *The Death and Life of Great American Cities*

Overview of the Field

Jacobs criticized both the Decentrists and the disciples of Le Corbusier. Howard's garden city assumed that a dense concentration of people was an inherent evil, whereas for Jacobs, dense concentration provided a source of vitality and creativity. Jacobs also dismissed Howard's vision of planned communities as "paternalistic, if not authoritarian."[3] While the Decentrists and the Le Corbusier school held different visions of the ideal community, both reflected a top-down approach. Le Corbusier's notion of Utopia was more libertarian; as Jacobs described it, "In his Radiant City nobody, presumably, was going to have to be his brother's keeper any more."[4] The Decentrists shaped the growth of the suburbs, while Le Corbusier's vision was most evident in New York City and Moses' planning.

As Jacobs could have used academic backup for her ideas, it is curious that she never addressed the Chicago School, known for its urban sociology.The scholars of the Chicago School, amongst whom were the founding urban sociologists Robert Park,* Louis Wirth,* and Ernest Burgess,* conducted their research at the University of Chicago during the 1920s and 1930s and are notable for their influence on the field of urban studies.*

Jacobs's lack of engagement with Park and Wirth is surprising, as social interaction within cities interested both men. Like Jacobs, Park had a background in journalism (though in Chicago) and encouraged his students to research urban life using an ethnographic (study of people and cultures) approach. Jacobs might have gleaned more insights if she had considered the work of the Chicago School, particularly Park's and Wirth's observations.

Academic Influences

Jane Jacobs was not an academic, and prevailing urban planning theories influenced her only in a negative sense. Bureaucratic failures compelled Jacobs to investigate what truly worked in cities. She credited a social worker with initiating the questions behind *Death and Life*: "The basic idea, to try to begin understanding the intricate social and economic order under the seeming disorder of the cities, was not my idea at all, but that of William Kirk, head worker of Union Settlement in East Harlem, New York, who, by showing me East Harlem, showed me a way of seeing other neighborhoods, and downtowns too."[5]

Jacobs also noted that Kirk called her attention to "the intricate social and economic order beneath the seeming disorder of cities."[6] Jacobs learned more about cities through observation than by studying theory. She rejected the authoritarian—dictatorial—nature of planners, insisting that communities and their residents knew best.

Because Jacobs based her analysis on observation, it makes sense that so many of her examples come from New York City and her Greenwich Village neighborhood. Yet other American cities followed New York's lead in urban renewal, with the same negative results. (Chicago's Eisenhower Expressway, for example, destroyed entire neighborhoods as it carved the nearby western suburb of Oak Park cleanly in half.) Jacobs explained that she first saw many of these trends in other cities: "In trying to explain the underlying order of cities, I use a preponderance of examples from New York because that is where I live. But most of the basic ideas in this book come from things I first noticed or was told in other cities."[7] Jacobs realized that New York City had become a "model" for other cities across the United States, where the growing list of failures derived from the same flawed assumptions. What the planners broke and could not then fix, she sought to set straight.

---

1 Jane Jacobs, *The Death and Life of Great American Cities* (New York: Vintage Books, 1992), 17.

2 Jacobs, *Death and Life*, 23.

3 Jacobs, *Death and Life*, 19.

4 Jacobs, *Death and Life*, 22.

5 Jacobs, *Death and Life*, 15–16.

6 Jacobs, *Death and Life*, 15.

7 Jacobs, *Death and Life*, 15.

MODULE 3
THE PROBLEM

KEY POINTS

- Jane Jacobs's main question was whether urban renewal helped or harmed cities. If it caused harm, she wanted to know which alternatives cities should pursue.
- When Jacobs wrote *The Death and Life of Great American Cities*, planners took the benefits of urban renewal for granted.
- Jacobs rejected the ideas of urban planning.* She insisted that observing social interaction was crucial for understanding cities.

Core Question

Jane Jacobs sought to answer one core question in *The Death and Life of Great American Cities*: was urban renewal good for cities—and if not, then what should cities do instead? To find answers, she had to tackle the larger issue of what makes cities work in the first place. Urban planners took these questions and their answers for granted. The biggest critics of planning were people living in the communities that urban renewal had devastated. Jacobs sided with their perspective, arguing, "Planners frequently seem to be less well equipped intellectually for respecting and understanding particulars than ordinary people, untrained in expertise, who are attached to a neighborhood, accustomed to using it, and so are not accustomed to thinking of it in generalized or abstract fashion."[1]

Jacobs had faith in a community's ability to understand cities better than planners—who never questioned the central tenets of urban planning the way Jacobs did. She asked critical questions about urban renewal that stemmed from neighborhood movements. Local protests in the 1950s fought the construction of the Cross Bronx Expressway,* a significant automobile route across the borough of New York known as the Bronx designed by the architect and planner Robert Moses.* After Jacobs read about Moses' plans to extend Fifth Avenue through the middle of Washington Square Park, she joined the Washington Square Park Committee,* founded by the community organizer Shirley Hayes.*[2] *Death and Life* found a wide audience because Jacobs gave voice to the growing sense of discontent about urban renewal—and instigated controversy because she challenged its basic assumptions.

> *"Jacobs derided urban renewal* as a process that served only to create instant slums. She questioned universally accepted articles of faith—for example that parks were good and that crowding was bad. Instead she suggested that parks were often dangerous and that crowded neighborhood sidewalks were the safest places for children to play."*
>
> —— Richard T. LeGates and Frederic Stout, *The City Reader*

The Participants

Jacobs attacked both the "city-destroying ideas"of the Decentrists and the urban planning of Le Corbusier.[3]

The Decentrists' ideas originated with Ebenezer Howard's model of the garden city,* and they favored a planning approach that stemmed from a dislike of cities and their dense concentrations of people. In Jacobs's time, the most influential disciple of the

Decentrists was Lewis Mumford,* an architecture critic for *The New Yorker** magazine. Mumford had written a book titled *The Culture of Cities*, which Jacobs derided as "a morbid and biased catalog of ills."[4] He responded with a review of *Death and Life* titled "Mother Jacobs Home Remedies."[5] Mumford's review acknowledged the significance of *Death and Life* and the novelty of its criticism, yet it had a condescending and even chauvinistic tone, as he labeled her approach "naked unawareness" and described it as "a mingling of sense and sentimentality, of mature judgments and schoolgirl howlers."[6]

Robert Moses never issued a public comment about *Death and Life*, though he returned the copy sent to him by its publisher, Random House, accompanied by a stern letter to the company's co-founder. "I am returning the book you sent me," Moses wrote. "Aside from the fact that it is intemperate and inaccurate, it is also libelous... Sell this junk to someone else."[7] *Death and Life* clearly touched a nerve in the man who considered himself New York's master builder.

The Contemporary Debate

When Jacobs wrote *Death and Life*, planners by and large did not question the wisdom of urban renewal—but her thinking about what makes cities great challenged their assumptions. Saskia Sassen,* a professor at Columbia University, argues that theorists had once seen the city as a "lens for understanding larger processes, a role it had lost by the 1950s."[8] However, in time, Jacobs's ideas gained greater acceptance as some planners began to incorporate her suggestions. In sum,"Jacobs's ideas have had a strong impact on the way urbanists and planners think about city life."[9] Following the publication of *Death and Life*, urban renewal came under greater scrutiny as planners and architects began to pay more attention to the dynamics of social interaction in cities.[10] In this way, Jacobs was truly ahead of her time.

Likewise, scholars who research cities and develop theories about them adopted more of Jacobs's ideas. *Death and Life* later became a foundational text for the school of thought known as New Urbanism,*[11] and scholars began to reconsider Jacobs's work in the light of gentrification,* the process whereby wealthy professionals migrate to central urban neighborhoods. Her ideas about mixed-use development,* aged buildings, and vibrant street life influenced the gentrification process—and as gentrification displaced working-class families, Jacobs's inattention to power and inequality came under scrutiny;[12] similarly, she failed to consider the ways in which race, ethnicity, class, and gender also shape how people experience the city. Yet her adversaries in the planning world never even bothered to consider people in the first place, and since then, these key questions of power and inequality have become central to urban studies.

---

1 Jane Jacobs, *The Death and Life of Great American Cities* (New York: Vintage Books, 1992), 441.

2 Anthony Flint, *Wrestling with Moses: How Jane Jacobs Took on New York's Master Builder and Transformed the American City* (New York: Random House, 2009), 75.

3 Jacobs, *Death and Life*, 17.

4 Jacobs, *Death and Life*, 20.

5 Lewis Mumford, "The Sky Line: 'Mother Jacobs Home Remedies,'" *The New Yorker*, December 1, 1962.

6 Kenneth Kidd, "Did Jane Jacobs' Critics Have a Point After All?" *The Toronto Star*, November 25, 2011, accessed August 31, 2015, http://www.thestar.com/news/ insight/2011/11/25/did_jane_jacobs_critics_have_a_point_after_all.html.

7 Flint, *Wrestling with Moses*, 125.

8 Saskia Sassen, "What Would Jane Jacobs See in the Global City, Place and Social Practices?" in *The Urban Wisdom of Jane Jacobs*, ed. Sonia Hirt and Diana Zahm (New York: Routledge, 2012), 84.

9 Mark Gottdiener, Ray Hutchinson, and Michael T. Ryan, *The New Urban Sociology*, 5th edn (Boulder, CO: Westview Press), 328.

10 Roberta Brandes Gratz, *The Battle for Gotham: New York in the Shadow of Robert Moses and Jane Jacobs* (New York: Nation Books, 2010).

11 Peter Katz, *The New Urbanism: Toward an Architecture of Community* (New York: McGraw-Hill Education, 1993).

12 Sharon Zukin, "Changing Landscapes of Power: Opulence and the Urge for Authenticity," *International Journal of Urban and Regional Research* 33, no. 2 (2009): 548–9.

MODULE 4
THE AUTHOR'S CONTRIBUTION

KEY POINTS

- Jane Jacobs aimed to show how cities work by investigating questions urban planners* had ignored.
- Urban planners assumed that cities were problematic and needed renewal. Sociologists from the Chicago School, an approach associated with the sociology* department at the University of Chicago, also maintained that cities were undesirable.
- Jacobs examined what made cities work because she could see that urban planning destroyed rather than renewed them.

Author's Aims

In *The Death and Life of American Cities*, Jane Jacobs seeks to understand and explain what makes cities great for the people who live in them. The work posed questions urban planners had failed to ask, largely because they assumed cities were undesirable places to live. Postwar policies supported a migration of the population into new suburbs, while planners promoted forms of urban renewal that devastated city neighborhoods.

Planners took an abstract view of cities that all but ignored the lives and needs of their inhabitants. Instead, they concerned themselves with the height of buildings, the creation of green spaces, and the efficient flow of traffic. Jacobs redefines their questions by returning the focus to how people used urban spaces. "I shall mainly be writing about common ordinary things," she writes at the beginning of *Death and Life*.[1] She then lists her concerns, which include:

- the safety of city streets
- the quality of city parks
- the condition of slums, and why they sometimes regenerated despite powerful opposition
- the shifting concerns of downtown districts
- the character and function of city neighborhoods[2]

Jacobs succeeded in shifting the terms of the debate about cities. She based her findings on observation and as a result reached vastly different conclusions to those of the urban planners.

> *"Jane Jacobs's ideas have influenced urbanists because she captured the heart and soul of urban culture. Her importance lies in convincing us that urban culture depends on the relationship between personal interaction and public space."*
>
> ——Mark Gottdiener, Ray Hutchison, and Michael T. Ryan, *The New Urban Sociology*

Approach

Jacobs broke new ground because she focused on how people interacted with and used city spaces. Her approach stood as the complete antithesis of urban planning's abstract theorizing. Urban planners saw little of value in city life, condemned entire neighborhoods as "slums," and set about destroying them. For them, the good life existed far from urban America, in small towns and suburbs. But Jacobs revealed how cities crackled with vitality, innovation, and diversity. "For Jacobs," in brief, "active urban life can never be planned because people invent uses for space."[3]

The inventiveness of *Death and Life* stemmed from Jacobs's methods for studying urban life. She watched how her neighbors went about their daily routines and observed that a city's social order "is kept primarily by an intricate, almost unconscious, network of voluntary controls and standards among the people themselves, and enforced by the people themselves."[4] In other words, social interaction on city streets creates a form of order that planners and architects cannot design.

In contrast, she saw how high-rise housing projects had become anonymous and dangerous. About such places she wrote,"No amount of police can enforce civilization where the normal,casual enforcement of it has broken down."[5] She also noted how the postwar suburbs produced monotony and demanded conformity—a view ahead of its time that wouldn't be embraced for decades. By focusing on the human scale of city life—and using original observational methods—Jacobs reached novel conclusions that flew in the face of urban planning's impersonal model.

Contribution in Context

Few scholars had studied urban life through routine social interaction, and so the originality of *Death and Life* came from Jacobs's street-level analysis, which flipped the top-down approach of urban planners on its head. While Jacobs did not engage with the Chicago School, known for urban sociology, its members had asked similar questions. The Chicago School focused more on urban subcultures than on the built environment of the city, with the sociologist Louis Wirth* of the University of Chicago more concerned with *urbanism* as a way of life than with *urbanization* as a physical transformation. For Wirth, urbanism's three characteristics were aggregate population size, density, and heterogeneity (which refers to its internal differences in things such as character, use, and population).[6]

Death and Life proved original because Jacobs not only studied cities at the street level but drew positive conclusions. Both planners and ethnographers—those studying a place's inhabitants—had assumed urban life was undesirable. Even urban sociologists of the Chicago School tended to view urbanism as a social ill, while planners in the years following World War II* advocated the depopulation of cities in favor of suburban living. Jacobs's aim, remarkable for its time, was to show how cities could foster both creativity and community. Figures such as Robert Moses* looked at city neighborhoods as chaotic slums begging for demolition and renewal. Jacobs argued that although city people were poor, these neighborhoods served as dynamic centers of interaction. Instead of renewing them, urban planning did quite the opposite, leaving countless corners of America's cities more anonymous, desolate, and alienating.

---

1 Jane Jacobs, *The Death and Life of Great American Cities* (New York: Vintage Books, 1992), 3–4.

2 Jacobs, *Death and Life*, 3–4.

3 Mark Gottdiener, Ray Hutchison, and Michael T. Ryan, *The New Urban Sociology*, 5th edn (Boulder, CO: Westview Press, 2015), 327.

4 Jacobs, *Death and Life*, 32.

5 Jacobs, *Death and Life*, 32.

6 Louis Wirth, "Urbanism as a Way of Life," *American Journal of Sociology*, 44, no. 1 (July 1938): 1–24.

SECTION 2
IDEAS

MODULE 5

MAIN IDEAS

KEY POINTS

- Jane Jacobs's main argument is that urban planners ruin cities because they fail to consider the city's dynamics of social interaction—the ways in which a city is defined by different social interactions in different contexts.
- Jacobs opposed the segregating, standardizing approach of urban planning.* She insisted that diversity, concentration, and mixture make cities great.
- Jacobs wrote *The Death and Life of Great American Cities* in direct, plain-spoken prose. This reflected her populist criticism of abstract urban theory.

Key Themes

The Death and Life of Great American Cities argues that urban planning destroys the vitality of city life. Jane Jacobs maintains that planners failed because they did not understand how social interaction creates great cities. While they presented urban renewal* to the public as a form of progress, Jacobs argued that it actually left cities worse off and more dangerous than before. Writing in 1961, she assesses what had become of cities in the wake of urban renewal: "Low-income projects that become worse centers of delinquency, vandalism, and general social hopelessness than the slums they were supposed to replace... Promenades go from no place to nowhere and have no promenaders. Expressways eviscerate great cities. This is not the rebuilding of cities.This is the sacking of cities."[1]

Jacobs develops four key themes in *Death and Life*:

- Urban renewal does not renew cities but actually destroys them.
- Planners ignore the everyday realities of social life in the city.
- Interaction of people on the streets makes cities "great."
- Great cities are built on diversity, concentration, and mixture instead of separation and standardization.

Because planners devised proposals and policies mainly from urban theory, they neglected to consider the micro-social dynamics of life in the city—the interactions between individuals. Jacobs saw this as a kind of futuristic fantasy that promised a more modern way of life but predictably resulted in failure and misery: "The pseudoscience of city planning and its companion, the art of city design, have not yet broken with the specious comforts of wishes, familiar superstitions, over simplifications, and symbols, and have not yet embarked upon the adventure of probing the real world."[2]

Instead of using powers of observation, bureaucrats (office-bound city officials) fixated on buildings and expressways, without ever taking into account the people affected by their construction.

> *"This book is an attack on current city planning and rebuilding. It is also, and mostly, an attempt to introduce new principles of city planning and rebuilding, different and even opposite from those now taught in everything from schools to architecture and planning to the Sunday supplements and women's magazines."*
>
> ——Jane Jacobs, *The Death and Life of Great American Cities*

Exploring the Ideas

Jacobs's damning criticisms reflected a different perspective gleaned through what she witnessed in her own neighborhood of New York's Greenwich Village. She insists that the life of any city lies in the commonplace actions and interactions of its millions of people—in which case urban planning can only succeed when its proposals take these social dynamics into account. Jacobs writes, "Cities have the capability of providing something for everybody, only because, and only when, they are created by everybody."[3] It is, she argues, entire communities of people engaging in interaction that creates great cities, and not a team of expert architects.

Yet these professionals and public officials lobbied for separation and standardization, which split commercial, residential, industrial, and cultural spaces. This led to homogeneous, sterile suburbs outside the city limits and anonymous, isolated housing projects within them. In contrast, Jacobs insists that diversity, concentration, and mixture made cities great. "To understand cities," Jacobs argues, "we have to deal outright with combinations or mixtures of uses, not separate uses, as the essential phenomena."[4] Against the prevailing wisdom, Jacobs contends that cities should incorporate spaces with mixed uses. This idea has been quite influential in contemporary urban planning,[5] while the conventions she fought against are today dismissed as incompatible with the creation of vibrant cities.

Diversity, concentration, and mixture also apply to people in the city and how they interact ("People gathered in concentrations of city size and density can be considered a positive good,"[6] Jacobs writes). Density maintains a strong connection to diversity, another facet of urban life that Jacobs valued but that planners of her era tended to dread. Jacobs maintains that the interaction between distinct kinds of people played an important role in creating vitality in cities, as she champions "a great and exuberant richness of differences and possibilities, many of these differences unique and unpredictable and all the more valuable because they are."[7]

Language and Expression

Jane Jacobs had little patience for the so-called experts who designed cities in abstraction from real life. She believed cities represented collections of ordinary people and commonplace actions. Likewise, she wrote *Death and Life* in plain, simple prose that avoids the jargon and cold language of urban planning and theory. Jacobs's concise, direct, and straightforward style jumps out right from the first sentence of *Death and Life*: "This book is an attack on current city planning and rebuilding."[8] The plain-spoken form of Jacobs's writing was a perfect match for the populist content of her critique.

Jacobs came to the study of cities and urban planning not as an academic, but as a journalist and New York City resident. This, along with the commonplace nature of her observations, gave *Death and Life* a degree of accessibility akin to that of a well-written magazine article. Jacobs began working as a journalist in the early 1940s, later writing for the journal *Architectural Forum*.* As Anthony Flint would write in his book *Wrestling With Moses*,

"Jacobs was a natural for all aspects of journalism and magazine publishing—a stickler for details, an authority on proper writing style and grammar, highly organized, and good at coming up with story ideas."[9] Jacobs brought her extraordinary talents as a journalist and a writer to bear in *Death and Life*, resulting in a devastating critique that brought down a planned expressway, and gave rise to a visionary view of cities.

---

1 Jane Jacobs, *The Death and Life of Great American Cities* (New York: Vintage Books, 1992), 4.

2 Jacobs, *Death and Life*, 13.

3 Jacobs, *Death and Life*, 238.

4 Jacobs, *Death and Life*, 144.

5 Roberta Brandes Gratz, *The Battle for Gotham: New York in the Shadow of Robert Moses and Jane Jacobs* (New York: Nation Books, 2010), 256–76.

6 Jacobs, *Death and Life*, 220–1.

7 Jacobs, *Death and Life*, 220–1.

8 Jacobs, *Death and Life*, 3.

9 Anthony Flint, *Wrestling with Moses: How Jane Jacobs Took on New York's Master Builder and Transformed the American City* (New York: Random House, 2009), 10.

MODULE 6
SECONDARY IDEAS

KEY POINTS

- Jane Jacobs offers several recommendations to supplement her critique of urban planning.* These consist of mixed-use development* (spaces where industrial, commercial, and cultural activities occur), short blocks, aged buildings, and dense concentration.
- As they embraced her critique, contemporary urban planners have adopted many of Jacobs's suggestions.
- Jacobs's advocacy of mixed-use spaces has had the largest impact on contemporary urban planning.

Other Ideas

Jane Jacobs's *The Death and Life of Great American Cities* primarily concerns itself with the effects of urban renewal.* In explaining why urban renewal failed, Jacobs investigates the opposite side of the coin: what makes cities work and what had once made them great.Where urban planners saw diversity as a source of chaos and disorder, Jacobs countered that it was a source of strength and vitality.

While *Death and Life* delivers a blistering critique of urban renewal, it offers smart alternatives; in the work's second part, Jacobs outlines four conditions that create diversity in cities:

- First, districts should include more than one primary function, thereby creating mixed-use spaces.
- Second, most blocks should be short, allowing more

opportunities to turn corners.

- Third, districts are best when buildings vary in age and condition.
- Fourth, there should be a sufficiently dense concentration of people.

Jacobs argues that all four conditions must be present to foster diversity, and so the absence of any one of them would limit the others. "The potentials of different districts differ for many reasons," she acknowledges, "but, given the development of these four conditions, a city district should be able to realize its best potential, wherever that may lie."[1] Not all cities are the same, but Jacobs believes that any of them could benefit from these four suggestions.

> *"Under the seeming disorder of the old city, wherever the old city is working successfully, is a marvelous order for maintaining the safety of the streets and the freedom of the city. It is a complex order. Its essence is intricacy of sidewalk use, bringing with it a succession of eyes."*
>
> ——Jane Jacobs, *The Death and Life of Great American Cities*

Exploring the Ideas

Urban planners preferred to separate residential, commercial, industrial, and cultural zones. However, mixed-use development, Jacobs contends, would ensure the circulation of people at various times of the day. Spaces with multiple uses attract "the presence of people who go outdoors on different schedules and are in the place for different purposes, but who are able to use many facilities in common."[2] When people occupy the streets at all times of the day, neighborhoods become safer and local businesses prosper.

Short blocks have a similar importance for stimulating diversity. The long streets and discrete neighborhoods urban planners created led to isolated, stagnant, and dangerous stretches. In contrast, Jacobs argues, "frequent streets and short blocks are valuable because of the fabric of intricate cross-use that they permit among the users of a city neighborhood." Short blocks enable the movement and mixture of pedestrians who use the same streets for different purposes.[3]

Jacobs's claim that cities need aged buildings went against both the grain of urban planning and the ideals of postwar modernism—which celebrated the new while disparaging the old in many forms of art and culture. Bureaucrats and their architects looked at neighborhoods with old buildings as blighted. However, while they promoted the process of demolition and renewal as progress, Jacobs counters, "If a city has only new buildings, the enterprises that can exist there are automatically limited to those that can support the high costs of construction."[4] She argues that preserving old buildings is essential to protect local businesses that can survive in them—and that would be crippled by the high overhead of new construction.

Jacobs's fourth condition of diversity involves dense concentrations of people. Once again, she goes against the prevailing ideas of the urban-planning establishment—after all, planners despised and feared population density. But Jacobs believes that density is one of the most important factors in making cities interesting and vital places—and insists that an abundance of people gives the city its unique character: "They should also be enjoyed as an asset and their presence celebrated: by raising their concentrations where it is needful for flourishing city life, and beyond that by aiming for a visibly lively public street life and for accommodating and encouraging, economically and visually, as much variety as possible."[5]

Overlooked

Many planners have since adopted Jacobs's recipe for fostering diversity. Her ideas, which once attracted the scorn of the planning establishment, have actually become part of mainstream planning. As a result, their original links with community organizing (as opposed to elite expertise) are now forgotten. Jacobs argues that communities know best, but this insight dissipated once planners started to utilize her ideas. In short, "Often overlooked is Jacobs' influence on community organizing... Jacobs' activist work showed people around the country that they could fight against the urban renewal bulldozer—and win."[6] Contemporary urbanists have credited Jacobs for changing the debate about how to plan cities. Yet they have neglected to recognize how her ideas emerged within a larger community engaged in a political struggle.

Another overlooked aspect of *Death and Life* concerns the relationship between women and the city. Writing in 1961, Jacobs's analysis predated by several years the emergence of second-wave feminism* (the social movement of women that began in the late 1960s, continuing earlier struggles for nondiscrimination, reproductive rights, and social equality). As the political scientist Marshall Berman* argued, "She makes her readers feel that women know what it is like to live in cities, street by street, day by day, far better than the men who plan and build them."[7] By focusing on the commonplace activities that occur on a daily basis, Jacobs shows herself as more attuned to a woman's perspective of city life. For Berman, she "gives us the first fully articulated woman's view of the city since [the social reformer] Jane Addams.*"[8]

Mainstream geographers and planners often bypass the feminism of Jacobs's critique because it preceded the second wave of feminism. Nevertheless, *Death and Life* features a feminist perspective on a crucial social and political topic that bears witness to how much a determined, visionary woman can achieve in the face of an entrenched male hierarchy.

---

1 Jane Jacobs, *The Death and Life of Great American Cities* (New York: Vintage Books, 1992), 151.

2 Jacobs, *Death and Life*, 152.

3 Jacobs, *Death and Life*, 186.

4 Jacobs, *Death and Life*, 187.

5 Jacobs, *Death and Life*, 221.

6 Peter Dreier, "Jane Jacobs' Radical Legacy," *National Housing Institute* 146 (Summer 2006), accessed September 5, 2015, http://www.nhi.org/online/ issues/146/janejacobslegacy.html.

7 Marshall Berman, *All That Is Solid Melts into Air: The Experience of Modernity* (New York: Penguin Books, 1982), 322.

8 Berman, *All That Is Solid Melts into Air*, 322.

MODULE 7

ACHIEVEMENT

KEY POINTS

- Jane Jacobs exposed and shattered the fundamental premises of the reconstructive policies of urban renewal* following World War II.*
- Jacobs's argument prevailed because she articulated popular dissatisfaction with urban renewal.
- Because she failed to consider economic factors alongside political ones, Jacobs left no basis for a critique of gentrification*—the process whereby wealthy professionals migrate to regenerating urban neighborhoods, pricing out the existing residents.

Assessing the Argument

Jane Jacobs's *The Death and Life of Great American Cities* is now regarded as a classic in urban studies,* and is often cited as a key text in the fields of urban sociology,* geography, and architecture.

Her recommendations are also frequently implemented in policy, as noted in a recent book, *The New Urban Sociology*: "Jacobs's ideas have had a strong impact on the way urbanists and planners think about city life. Local governments encourage park use, street festivals, temporary blocking of community roads, and toleration of sidewalk vendors."[1] Indeed, few books have made so many contributions to theory and practice in such a number of disciplines. The publication of *Death and Life* marked a turning point for urban studies.

Still, not all the ideas in *Death and Life* have proven acceptable or successful. As the coauthors of *The New Urban Sociology* note,"Some of her followers advocated the elimination of elevators in apartment buildings to facilitate neighborly interaction, but the results were disastrous for the residents of these buildings."[2] One crucial deterrent to the implementation of Jacobs's ideas is crime. So it was that, in many cities,"downtown revitalization efforts using Jacobs's ideas have failed due to the fear of urban crime on the part of suburban residents."[3] Finally, permeating *Death and Life* is a strong streak of sentimentality and nostalgia that no longer pertains to city life. In sum, "Jacobs's ideas about community may also be passé. Many city residents socialize with networks of friends and relatives who do not live nearby... Teenagers may prefer to travel to their own friendship networks rather than socialize on the street."[4]

> *"The great urban theorist Jane Jacobs was not an academically trained economist, but her theory of growth made an indelible contribution to the field. In her eyes, it was new types of work and new ways of doing things that drove large-scale economic expansions. But while most economists located momentum in great companies, entrepreneurs, and nation-states, Jacobs identified great cities as the prime motor force behind innovation."*
>
> ——Richard Florida, *The Rise of the Creative Class*

Achievement in Context

Jacobs wrote *Death and Life* at a time when communities began to challenge urban renewal. These social circumstances bolstered the book's success. Neighborhood movements arose in the New

York borough of the Bronx during the 1950s in opposition to the building of a major road, the Cross Bronx Expressway,* and *Death and Life* appeared in 1961 as residents fought the architect and planner Robert Moses'* plans for urban renewal in Lower Manhattan. *Death and Life* articulated a new way of thinking about cities: "It became an inspiration, guidebook, and bible for not only a new generation of planners and architects but ordinary citizens as well."[5] These external events of political conflict made *Death and Life* one of the most urgent books of its time.

Death and Life was the first of several books to impact social issues and public policy in the early 1960s. Other books published shortly afterwards that roused public awareness included the marine biologist and environmentalist Rachel Carson's* *Silent Spring* (1962); the social campaigner Michael Harrington's* *The Other America* (1962); the feminist author Betty Friedan's* *The Feminine Mystique* (1963); and the activist and consumer advocate Ralph Nader's* *Unsafe at Any Speed* (1965). Carson's *Silent Spring* helped launch environmentalism with its exposé of pesticide use.[6] Harrington's *The Other America* provided a catalyst for the War on Poverty.[7] *The Feminine Mystique* became a foundational text for second-wave feminism; Friedan would establish the National Organization for Women (NOW)* in 1966.[8] And Nader's *Unsafe at Any Speed* exposed the auto industry's evasion of safety precautions (using the American car manufacturer Chevrolet's Corvair as a focal point) and provided the impetus for consumer advocacy.[9]

Death and Life also offered a rationale for the "back to the city" movement that anticipated gentrification. Starting in the 1950s, middle-class professionals in New York began relocating to formerly working-class neighborhoods in the borough of Brooklyn such as Brooklyn Heights, Park Slope, and Boerum Hill, where they renovated period houses.[10] Jacobs's book became an influential text for them following its publication: "Most of the middle-class enthusiasts inspired to move to Brooklyn by Jane Jacobs' *Death and Life of Great American Cities* cited her sentimental 'street ballet' or the passages where she slipped into simple romantic nostalgia."[11] An unanticipated effect of *Death and Life* was its stimulation of gentrification—which contemporary critics have blamed for displacing the working class and the poor, people of color, and ethnic communities from urban neighborhoods.

Limitations

In the decades following the publication of *Death and Life*, scholars have identified at least two major shortcomings. First, Jacobs does not pay attention to the economic factors that drive urban development. Although she distrusted the state power embodied by Robert Moses, capitalist developers escaped her critique. And so, while "Jacobs and other [residents of the neighborhood of Greenwich Village] were busy fighting off Moses' large-scale public works plans, they failed to oppose the quieter, piecemeal incursions of private real estate developers, who through the 1950s were demolishing older buildings all over the Village for new high-rises."[12] Though *Death and Life* presented a devastating assessment of urban renewal, Jacobs's inattention to capitalist development ensured her complicity with gentrification.

A second limitation concerns Jacobs's nostalgic ideal of great cities, in which older neighborhoods are seen as being more close-knit and down to earth than their suburban successors. This derives from the authenticity of ethnic neighborhoods that Jacobs imagines. In the years after *Death and Life*, gentrifying professionals returned to the city in search of this authenticity. In brief, "Jacobs fails to recognize the growing influence of her own perspective, to see that families like hers are gradually moving to the West Village's nineteenth century houses because they appreciate the charm of the area's little shops and cobblestone streets."[13] In other words, Jacobs idealized older cities and urban communities in ways later emulated by young, white-collar professionals who gentrified American cities such as Chicago and San Francisco. The irony of gentrification is that it uproots the ethnic and artistic enclaves that professionals picture as the embodiment of authenticity.

---

1 Mark Gottdiener, Ray Hutchison, and Michael T. Ryan, *The New Urban Sociology*, 5th edn (Boulder, CO: Westview Press, 2015), 328.

2 Gottdiener, Hutchison, and Ryan, *The New Urban Sociology*, 328.

3 Gottdiener, Hutchison, and Ryan, *The New Urban Sociology*, 328.

4 Gottdiener, Hutchison, and Ryan, *The New Urban Sociology*, 328.

5 Anthony Flint, *Wrestling with Moses: How Jane Jacobs Took on New York's Master Builder and Transformed the American City* (New York: Random House, 2009), 129.

6 Rachel Carson, *Silent Spring* (Boston: Houghton Mifflin Company, 1962).

7 Michael Harrington, *The Other America: Poverty in the United States* (New York: Touchstone, 1962).

8 Betty Friedan, *The Feminine Mystique* (New York: W. W. Norton, 1963).

9 Ralph Nader, *Unsafe at Any Speed: The Designed-In Dangers of the American Automobile* (New York: Grossman, 1965).

10 Suleiman Osman, *The Invention of Brownstone Brooklyn: Gentrification and the Search for Authenticity in Postwar New York* (New York: Oxford University Press, 2011).

11 Osman, *The Invention of Brownstone Brooklyn*, 169.

12 John Strausbaugh, *The Village: A History of Greenwich Village* (New York: Harper Collins, 2013), 341.

13 Sharon Zukin, *Naked City: The Death and Life of Authentic Urban Places* (New York: Oxford University Press, 2010), 18.

MODULE 8

PLACE IN THE AUTHOR'S WORK

KEY POINTS

- Jane Jacobs had a wide range of interests, but throughout her life she focused on cities and their economies.
- Although she wrote six more books, *The Death and Life of Great of American Cities* remained her most well known.
- *Death and Life* established Jacobs's international reputation as an authority on cities and urban living.

Positioning

The Death and Life of Great American Cities certainly qualifies as Jane Jacobs's best-known work. She had outlined some of the book's major themes in an earlier article, "*Downtown Is for People*," published by *Fortune* magazine in 1958. Jacobs had given a talk for the Conference on Urban Design at Harvard University's Graduate School of Design, and *Fortune* writer and editor William H.Whyte* asked her to develop it into an article. In "*Downtown Is for People*," Jacobs presented her main critique of urban renewal: "These projects will not revitalize downtown; they will deaden it...They will be stable and symmetrical and orderly. They will be clean, impressive, and monumental. They will have all the attributes of a well-kept, dignified cemetery."[1] The *Fortune* article stated some of her arguments against urban renewal,* but she had yet to define what made cities work.

Death and Life was Jacobs's first of seven books, most of which focused on cities and urban development. She began writing her second book, *The Economy of Cities*, during her years of activism in New York, and completed it shortly after moving to Toronto. In a chapter titled "The Valuable Inefficiencies and Impracticalities of Cities," Jacobs compared the economies of the English cities Manchester and Birmingham. While Manchester teetered on the brink of becoming an "obsolescent city," Birmingham's "fragmented and inefficient little industries kept adding new work, and splitting off new organizations, some of which became very large but were still outweighed in total employment and production by the many small ones." *The Economy of Cities* continued to develop a central conclusion she had reached in *Death and Life*: diversity was crucial to the life of cities, while large-scale projects were counterproductive.[2]

> *"None of [her other] books were blockbusters like* The Death and Life of Great American Cities*, and Jacobs began to chafe when the questions inexorably led back to her days among the bohemians in Greenwich Village, fighting the New York battles—as if she were a rock star constantly being asked to play an old hit."*
>
> ——Anthony Flint, *Wrestling with Moses: How Jane Jacobs Took on New York's Master Builder and Transformed the American City*

Integration

The rest of Jacobs's books explored a range of other concerns, often departing from the analysis of cities and their economies. Her third book, 1980's *The Question of Separatism*, made a case for the French-speaking region of Quebec's independence from Canada.[3] Although this book digressed from her concern with cities, it continued the advocacy of popular self-determination that began with *Death and Life*. Her next book, *Cities and the Wealth of Nations*, revisited the topic of urban economics, but her fifth, *Systems of Survival*, explored new questions of morality and values.[4] *Systems of Survival* examined the different moral precepts behind what Jacobs labeled a "guardian syndrome" and a "commerce syndrome." She returned to the study of economics in *The Nature of Economies*, but her final book, *Dark Age Ahead*, was an apocalyptic warning about the decay of North American society.[5]

Although Jacobs maintained a lifelong interest in cities and economics, her books did not center on a single, coherent idea. Instead, she explored a wide range of questions and issues. Yet for all her productivity, *Death and Life* remained her most influential book. Although Jacobs saw herself as a scholar with broad concerns, the public associated her with cities and the opposition to urban renewal in the 1960s.

Significance

While her other books were largely forgotten, *Death and Life* has proved to be a lasting influence. It established Jacobs's reputation as a leading expert on cities, but it also inspired as much as it informed. As the author Anthony Flint writes in his book about Jacobs's battle for a new kind of city regeneration, *Wrestling with Moses*, "The book's influence was undeniable for a new generation of citizen activists, students—who viewed her as a kind of folk hero—and city planners. Activists in cities across the United States modeled themselves after Jacobs, acting as watchdogs over local government and demanding to be heard on everything from street-corner wastebaskets to the shadows cast by proposed skyscrapers."[6]

Death and Life's enduring impact has now lasted more than half a century and its influence has increased over time. In the book, Jacobs predicted the changes cities would undergo after the 1960s. For example, white-collar, gentrifying professionals now find cities desirable places to live for many of the same reasons that Jacobs first outlined. Her ideas are also realized in forms of postmodern architecture* that, starting in the 1980s, became more pervasive (in architecture,"postmodern" refers to a movement that sought to move on from the dominant style of postwar architecture, notably the architectural style known as "modernism"). As the urban theorist David Harvey* wrote,"Architecture and urban design have therefore been presented with new and more wide-ranging opportunities to diversify spatial form than was the case in the immediate postwar period. Dispersed, decentralized, and deconcentrated urban forms are now much more technologically feasible than they once were."[7]

In this way, Jacobs's ideas, which began as a grassroots rallying cry to stay one step ahead of the bulldozer, proved to be ahead of their time as well.

1 Jane Jacobs, "Downtown Is for People," *Fortune* (1958), accessed September 6, 2015, http://fortune.com/2011/09/18/downtown-is-for-people-fortune-classic-1958/.

2 Jane Jacobs, *The Economy of Cities* (New York: Random House, 1969), 88–9.

3 Jane Jacobs, *The Question of Separatism: Quebec and the Struggle Over Sovereignty* (New York: Random House, 1980).

4 Jane Jacobs, *Cities and the Wealth of Nations* (New York: Vintage Books, 1984); Jane Jacobs, *Systems of Survival* (New York: Random House, 1992).

5 Jane Jacobs, *The Nature of Economies* (New York: Random House, 2000); Jane Jacobs, *Dark Age Ahead* (New York: Random House, 2004).

6 Anthony Flint, *Wrestling with Moses: How Jane Jacobs Took on New York's Master Builder and Transformed the American City* (New York: Random House, 2009), 185.

7 David Harvey, *The Condition of Postmodernity* (Cambridge, MA: Blackwell, 1990), 75–6.

SECTION 3
IMPACT

MODULE 9

THE FIRST RESPONSES

KEY POINTS

- Jane Jacobs's critics derided her for offering unrealistic alternatives to urban renewal* and ignoring factors such as social class.
- Jacobs did not directly respond to criticism of her ideas but did object to the condescending tone of many of her critics.
- The devastating effects of urban renewal in many cities shaped a consensus that *The Death and Life of Great American Cities* was the right book for its time.

Criticism

Jane Jacobs's *The Death and Life of Great American Cities* received both accolades and criticism following its publication. Writing in the *New York Times* a few months after the book's publication, the notable urban studies* scholar Lloyd Rodwin* offered a dose of both. "Readers will vehemently agree or disagree with the views," Rodwin wrote,"but few of them will go through the volume without looking at their streets and neighborhoods a little differently, a little more sensitively."[1] Indeed, it was impossible to ignore *Death and Life*'s originality or the significance of its impact. The editors of *Architectural Forum*,* for whom Jacobs had once written, were even more enthusiastic: "Is it not wonderful whenever long-accepted notions in *any* field are challenged, especially when that challenge is made with high intelligence and on humanistic grounds?"[2] Love it or hate it, *Death and Life* left no doubt that Jane Jacobs had issued a major challenge to the dominant ideas of urban planning.*

One significant critic of *Death and Life* was Lewis Mumford,* a historian of cities who wrote a regular column about architecture for *The New Yorker** magazine. Mumford, who had also opposed some of Robert Moses'* urban renewal projects, encouraged Jacobs after hearing her talk at Harvard's Graduate School of Design and reading her earlier article in *Fortune*, "Downtown Is for People." Yet Jacobs had criticized Mumford's book *The Culture of Cities* in her book, along with prominent urban planners—and Mumford countered with a condescending review of *Death and Life* titled "Mother Jacobs Home Remedies."[3] He dismissed Jacobs as an academic dilettante, an amateur: "I shall say no more of Mrs. Jacobs's lack of historical knowledge and scholarly scruple except that her disregard of easily ascertainable facts is all too frequent."[4]

A more balanced, respectful review came from sociologist Herbert Gans* in *Commentary* magazine. Gans praised *Death and Life* as "a thoughtful and imaginative tract on behalf of the traditional city."[5] But he also took issue with what he called Jacobs's "physical fallacy," which led her to "ignore the social, cultural, and economic factors that contribute to vitality or dullness."[6] In brief, Jacobs's analysis failed to consider additional variables such as social class. "In proposing that cities be planned to stimulate an abundant street life," he continued, "Jacobs not only overestimates the power of planning in shaping behavior, but she in effect demands that middle-class people adopt working-class styles of family life, child rearing, and sociability."[7] Gans argued that this perspective limited the scope of her analysis.

> *"*Death and Life *never made the bestsellers list, but as soon as it came out, it hit a nerve. Everywhere, people were talking about it, in newspaper editorials and book reviews, in classrooms and boardrooms, and in public symposia. Jacobs had stated her case so forcefully that you couldn't ignore* Death and Life *even if you hated it, as, predictably, many did—especially politicians and developers, and anybody else whose interests she had attacked."*
>
> ——Alice Sparberg Alexiou, *Jane Jacobs: Urban Visionary*

Responses

For the most part, Jacobs did not directly reply to her critics; but she continued to take part in activism against urban renewal, a strong indication that she was willing to stand by her argument. Jacobs further developed many of the ideas initially laid out in *Death and Life* in her second book, *The Economy of Cities*.

More than four decades after *Death and Life*'s publication, Jacobs recounted her feelings about Mumford's review in a 2004 lecture at City College in New York. She ridiculed Mumford's smug sexism, which other critics also employed:"I thought his reaction to the book was not quite rational... Maybe if he'd lived at a different time he would have understood that women didn't necessarily aspire to be patronized. He believed that women were a sort of ladies' auxiliary of the human race."[8]

Still, Jacobs seemed hurt by her more thoughtful and less politically motivated critics. Her biographer noted that "she often does not take criticism well."[9] In particular, she was "very angry" with Herbert Gans's review in *Commentary*, which balanced praise and criticism.[10] Jacobs and Gans had previously met and exchanged ideas when she was in Boston, and in 1962 Gans focused in his own book, *The Urban Villagers*, on the struggle with urban renewal experienced by Boston's North End neighborhood.[11]

Conflict and Consensus

The general consensus among Jacobs's critics was that she correctly identified the negative impact of urban renewal. Indeed, its devastating effects were evident in any number of cities across the country. Yet many also agreed that her solutions and suggestions were misguided or simply wrong. In short, many held that "it was Jacobs's analysis that made her book brilliant, but not her prescriptions."[12] As Gans pointed out in his review of *Death and Life*, most middle-class people did not necessarily desire diversity, were more drawn to suburban neighborhoods than to urban ones, and were not eager to walk or use public transportation.[13]

Over time, debates about *Death and Life* shifted in focus. As American politics took a turn to the right, some leftist critics observed that conservatives could appropriate her ideas. In 1982, the political scientist Marshall Berman* wrote, "What is relevant and disturbing here is that ideologues of the New Right have more than once cited Jacobs as one of their patron saints... It seems to me that beneath her modernist text there is an anti-modernist subtext, a sort of undertow of nostalgia for a family and a neighborhood in which the self could be securely embedded."[14]

The nostalgic strain of *Death and Life* began to mesh with conservative values and sentimentality in the 1980s. This had not necessarily been Jacobs's intent, but the shifting social context produced some apparent common ground—unusually, since conservatives of Jacobs's time would hardly have considered themselves defenders of the city.

---

1 Lloyd Rodwin, "Neighbors Are Needed," *New York Times*, November 5, 1961.

2 Quoted in Alice Sparberg Alexiou, *Jane Jacobs: Urban Visionary* (New Brunswick, NJ: Rutgers University Press, 2006), 84.

3 Lewis Mumford, "The Sky Line: 'Mother Jacobs Home Remedies,'" *The New Yorker*, December 1, 1962.

4 Mumford, "The Sky Line."

5 Herbert Gans, "City Planning and Urban Realities," *Commentary* 33 (February 1962): 170–5.

6 Gans, "City Planning and Urban Realities."

7 Gans, "City Planning and Urban Realities."

8 Quoted in Alexiou, *Jane Jacobs*, 94.

9 Alexiou, *Jane Jacobs*, 94.

10 Alexiou, *Jane Jacobs*, 94.

11 Herbert Gans, *The Urban Villagers: Group and Class in the Life of Italian-Americans* (New York: Free Press of Glencoe, 1962).

12 Alexiou, *Jane Jacobs*, 83.

13 Gans, "City Planning and Urban Realities."

14 Marshall Berman, *All That Is Solid Melts into Air: The Experience of Modernity* (New York: Penguin Books, 1982), 320.

MODULE 10
THE EVOLVING DEBATE

KEY POINTS

- Jane Jacobs's ideas gave new relevance to the study of gentrification—a process in which a neighborhood is regenerated so successfully that its original residents can no longer afford it—and influenced alternative forms of urban planning.*
- The New Urbanism* school of thought was inspired in part by *The Death and Life of American Cities*.
- Jacobs's ideas were transformative for urban studies* and many of its leading scholars, such as Richard Sennett* and Sharon Zukin.*

Uses and Problems

In *The Death and Life of Great American Cities* Jane Jacobs attacked ideas of urban planning that had been dominant since the end of World War II.* By the time of the book's publication in 1961, the urban renewal* projects of planners such as Robert Moses* were coming to an end. Cities faced an era of change, and Jacobs's ideas both indicated and catalyzed those changes. The first signs of gentrification had begun to appear, transforming many urban neighborhoods by the 1980s. In retrospect, Jacobs's ideas acquired renewed importance as a portent of those changes: "Gentrification was the central tension in her work, but in 1961 Jacobs did not yet have the vocabulary to make sense of it. Her Hudson Street in reality was not a quaint village but a dynamic middle ground lodged fragilely between an expanding postindustrial landscape and a declining industrial one."[1]

At the leading edge of gentrification were young middle-class professionals in search of an authentic feel and lifestyle they could not find in the suburbs. They associated this with the ethnic communities, local businesses, aged buildings, and walkable streets that Jacobs described in *Death and Life*. As the urban sociologist and leading theorist of gentrification Sharon Zukin put it, "Jane Jacobs expressed the appeal of this new sense of urban authenticity better than anyone."[2] *Death and Life* has taken on new significance as urban theorists seek to understand gentrification—and the white-collar quest for the authentic that fuels it.

> *"If writers like Jane Jacobs sought to rescue place from modern capitalist assimilation, they also commemorated a specific locale in the metropolis at a fleeting moment in the city's evolution: a nineteenth-century industrial middle cityscape on the periphery of the postwar modern central business district or university campus, in the early stages of gentrification.*"*
>
> ——Suleiman Osman, *The Invention of Brownstone Brooklyn*

Schools of Thought

Jacobs became one of the most important influences in the rise of the design and planning movement known as New Urbanism. This loosely defined group includes architects, planners, and scholars who advocate planning diversity, stress the importance of street life, and offer specific architectural solutions to address these goals.[3] Their core principles were laid out in a 1993 charter: "The

Congress for the New Urbanism views disinvestment in central cities, the spread of placeless sprawl, increasing separation by race and income, environmental deterioration, loss of agricultural lands and wilderness, and the erosion of society's built heritage as one interrelated community-building challenge."[4]

New Urbanism built on Jacobs's ideas and applied them to the contemporary problems of social inequality and ecological peril.

Since the 1990s, New Urbanism has attracted many practitioners and has become one of the most influential movements in urban planning. Its adherents have also incited no shortage of criticism. For example, the urban theorist David Harvey*—a scholar who addresses the relationship between capitalism and social space, an approach typical of those who draw on the theoretical tools provided by Marxist* analysis of society and economics—accused the New Urbanists of mainly addressing the needs of young middle-class professionals, while ignoring the plight of the inner city's underclass.[5] Others have criticized New Urbanism's disciples because they apply their ideas to smaller cities, planned communities, and neighborhoods not integrated into the metropolis; some also argue that their projects feel artificial and too strictly planned.[6]

Jacobs herself seemed to echo these criticisms when she was asked about New Urbanism in a 2001 interview: "The New Urbanists want to have lively centers in the places that they develop, where people run into each other doing errands and that sort of thing. And yet, from what I've seen of their plans and the places they have built, they don't seem to have a sense of the anatomy of these hearts, these centers. They've placed them as if they were shopping centers. They don't connect."[7] Thus, although Jacobs clearly inspired the New Urbanists, she maintained that they misinterpreted her ideas.

In Current Scholarship

Jacobs's *Death and Life* has impacted many contemporary scholars and practitioners of urban design and planning. Jacobs has influenced two leading American urbanists, Richard Sennett and Sharon Zukin, though they also maintain a critical distance from some of her ideas. Sennett is interested in how people experience cities through their bodily senses. He uses methods similar to Jacobs's to examine everyday life and social interaction in Greenwich Village. "Like so many others," Sennett wrote, "I had read my way into Greenwich Village, before arriving there twenty years ago, in the pages of Jane Jacobs's *The Death and Life of Great American Cities*."[8] While Sennett ultimately disagreed with many of Jacobs's conclusions, his intellectual debt to her is clear.

Sharon Zukin is another scholar indebted to Jacobs, even though she also diverged from her ideas. In her studies of gentrification, Zukin further developed some of Jacobs's ideas about the intersections of cities and culture. She has also examined issues Jacobs neglected, especially questions of race, class, and social inequality in the city.Yet while Zukin differentiated herself from Jacobs, she recognized that *Death and Life* had continuing relevance for the study of gentrification, and in fact had anticipated this change: "She connected small, old buildings and cheap rents with neighborhood street life, specialized, low-price shops, and new, interesting economic activities: in other words, downtown's social values."[9]

---

1 Suleiman Osman, *The Invention of Brownstone Brooklyn: Gentrification and the Search for Authenticity in Postwar New York* (New York: Oxford University Press, 2011), 177.

2 Sharon Zukin, *Naked City: The Death and Life of Authentic Urban Spaces* (New York: Oxford University Press, 2010), 16.

3 Peter Katz, *The New Urbanism: Toward an Architecture of Community* (New York: McGraw-Hill, 1994).

4 Congress for the New Urbanism, "Charter of the New Urbanism," in *The City Reader*, 5th edn, ed. Richard T. LeGates and Frederic Stout (New York: Routledge, 2011), 357.

5 David Harvey, "The New Urbanism and the Communitarian Trap," *Harvard Design Magazine* 1 (1997): 68–9.

6 See Todd W. Bressi, ed., *The Seaside Debates: A Critique of the New Urbanism* (New York: Rizzoli, 2002).

7 Jane Jacobs and Bill Steigerwald, "City Views: Urban Studies Legend Jane Jacobs on Gentrification, the New Urbanism, and Her Legacy," Reason 33, no. 2 (June 2001).

8 Richard Sennett, *Flesh and Stone: The Body and the City in Western Civilization* (New York: W. W. Norton, 1994), 355.

9 Sharon Zukin, *Landscapes of Power: From Detroit to Disney World* (Berkeley and Los Angeles: University of California Press, 1991), 191.

MODULE 11

IMPACT AND INFLUENCE TODAY

KEY POINTS

- *The Death and Life of Great American Cities* is widely considered a classic work of urban studies* and urban planning.*
- *Death and Life* continues to challenge urban planners to build cities that improve the quality of life for their inhabitants.
- The debates continue between theorists and planners concerning the rightful heirs to Jacobs's ideas.

Position

Jane Jacobs's *The Death and Life of Great American Cities* is indisputably an urban studies classic. The general consensus is that the book transformed approaches to the city in both theory and practice. *Death and Life* opened new theoretical debates about the function of cities in postwar America and abroad, and marked a significant change in how cities were designed and planned. Jacobs's book regularly appears on reading lists for courses in a broad array of disciplines and is often reprinted in excerpts for edited anthologies about cities;[1] and it is almost always discussed in urban studies textbooks.[2]

In the decades since *Death and Life*'s publication, critiques have shifted from institutional questions about urban renewal* to Jacobs's neglect of issues such as race, class, and social inequality. The sociologist Sharon Zukin* and the Marxist* geographer David Harvey* have stood at the forefront of this changing focus. Jacobs was initially criticized for her lack of academic qualifications, absence of a systematic methodology, and oversimplification of the literature.[3] However, these objections are now seen as less important, whereas her disregard of race and class poses more problems. Her idealization of Greenwich Village, for example, lacks any analysis of inequality:"This is what makes her neighborhood vision seem pastoral: it is the city before the blacks got there. Her world ranges from solid working-class whites at the bottom to professional middle-class whites at the top."[4] So while *Death and Life* is still considered a classic, scholars interested in race and class must look elsewhere.

> *"The girl from Scranton stood up to Moses* and challenged the status quo. Now virtually all those engaged in city building follow her rules. Her triumphs are engraved in the protocols followed by developers, city officials, and advocacy and grassroots organizations, and copies of* The Death and Life of Great American Cities *sit on the shelves of the planning offices at city halls across the country."*
>
> ——Anthony Flint, ***Wrestling with Moses: How Jane Jacobs Took on New York's Master Builder and Transformed the American City***

Interaction

Jacobs wrote *Death and Life* in a historical context in which urban renewal was ravaging cities, and where a litany of faulty assumptions shaped the approach of planners. When Jacobs challenged this orthodoxy,* it lost its unquestioned power. Since then, many of the core ideas of *Death and Life* have become the new common sense of urban design and planning. Therefore, as the debates within urban studies have long since moved on, *Death and Life*'s critique of city planning has become canonical.

One group of theorists and planners who have inherited and extended Jacobs's ideas, including Richard Sennett* and the urban studies professor Richard Burdett* of the London School of Economics, advocate for an "open city"*—a city based on principles of democracy and diversity. Sennett's use of Jacobs follows the spirit and intent of *Death and Life*; he invokes Jacobs in explaining the open-city idea. He argues that Jacobs "believes that in an open city, as in the natural world, social and visual forms mutate through chance variation; people can best absorb, participate, and adapt to change" in what he calls "urban time."[5]

New Urbanism* is composed of a second group of theorists and planners whom Jacobs influenced. In their planned communities, the New Urbanists have incorporated many of Jacobs's suggestions for mixed-use districts,* short blocks, and pedestrian-friendly neighborhoods. These core principles are outlined in their founding charter: "Neighborhoods should be diverse in use and population; communities should be designed for the pedestrian and transit as well as the car; cities and towns should be shaped by physically defined and universally accessible public spaces and community institutions; urban places should be framed by architecture and landscape design that celebrate local history, climate, ecology, and building practice."[6]

Although Jacobs distanced herself from New Urbanism,[7] *Death and Life* is clearly a foundational text for its ideas.

The Continuing Debate

Jacobs has become a venerated figure in urban studies. Yet while numerous theorists and planners claim her as an influence, some points of debate remain. One contemporary scholar who challenges Jacobs and her disciplines is the Harvard economist Edward Glaeser.* In his book *Triumph of the City*, Glaeser declares, "Many of the ideas in this book draw on the wisdom of Jane Jacobs, who knew that you need to walk a city's streets to see its soul."[8] Nevertheless, Glaeser proceeds to argue against some of Jacobs's conclusions on theoretical and methodological grounds. Jacobs, he maintains, "also made mistakes that came from relying too much on her ground-level view and not using conceptual tools that help one think through an entire system."[9] Specifically, Glaeser challenges her conclusion that restricting the height of buildings and preserving older neighborhoods would make cities more affordable—and suggests that planners should revisit the benefits of high-rise apartments and skyscrapers.

Like advocates of the open city and New Urbanists, Glaeser is concerned with the environmental impact of urban living, but less troubled by its social inequalities. The varying responses to Jacobs are partly a result of the different ways of putting her ideas into practice. Glaeser is an urban economist who examines cities as a scholar and theorist; the New Urbanists, by contrast, are mostly practitioners who embody her notions in planned communities and architecture. Glaeser claims to work in the tradition of Jacobs, as do the advocates of the open city and the planners who practice New

Urbanism—but to be sure, the conclusions each of these disciples reach are very different.

---

1 See Richard T. LeGates and Frederic Stout, eds, *The City Reader*, 5th edn (New York: Routledge, 2011); Janet Lin and Christopher Mele, eds, *The Urban Sociology Reader*, 2nd edn (New York: Routledge, 2012).

2 See Mark Gottdiener, Ray Hutchison, and Michael T. Ryan, *The New Urban Sociology*, 5th edn (Boulder, CO: Westview Press, 2015); John J. Macionis and Vincent M. Parrillo, *Cities and Urban Life*, 6th edn (Boston, MA: Pearson, 2013).

3 See Lewis Mumford, "The Sky Line: 'Mother Jacobs Home Remedies,'" *The New Yorker*, December 1, 1962.

4 Marshall Berman, *All That Is Solid Melts into Air: The Experience of Modernity* (New York: Penguin Books, 1982), 324.

5 Richard Sennett, "The Open City," *Urban Age* (2006): 2–3.

6 Congress for the New Urbanism, "Charter of the New Urbanism," in *The City Reader*, 5th edn, ed. Richard T. LeGates and Frederic Stout (New York: Routledge, 2011), 357.

7 Jane Jacobs and Bill Steigerwald, "City Views: Urban Studies Legend Jane Jacobs on Gentrification, the New Urbanism, and Her Legacy," *Reason* 33, no. 2 (2001): 48–55.

8 Edward Glaeser, *Triumph of the City: How Our Greatest Invention Makes Us Richer, Smarter, Greener, Healthier, and Happier* (New York: Penguin Press, 2011), 11.

9 Glaeser, *Triumph of the City*, 11.

MODULE 12
WHERE NEXT?

KEY POINTS

- As the populations of cities around the world continue to increase, *The Death and Life of American Cities* is likely to grow in influence.
- *Death and Life* will continue to serve as a foundational text for theories of urban planning* that supplanted the ideas and practices typical of the years following World War II.*
- *Death and Life* was a groundbreaking work in its critique of urban renewal,* its description of social interaction in cities, and its foreshadowing of gentrification.*

Potential

Jane Jacobs's *The Death and Life of Great American Cities* could be about to exert its influence on theorists and planners outside the United States. As of 2015, a majority of the world's people live in urban areas—a first in global history.[1] Cities around the world are growing faster than ever before, with no end in sight. The United Nations estimates that if growth continues at the current rate, the world's urban population will double every 38 years.[2] Clearly, these conditions demand a sharper understanding of what makes cities work and how to plan them, and this is where Jacobs's book could be influential.

Global cities can benefit from incorporating Jane Jacobs's insights about the relationship between economics and space. Contemporary urban theorists such as Richard Florida* emphasize the importance of cities for economic innovation.[3] Florida and many others—including those outside the academic realm—recognize that urban spaces play a crucial role in nurturing the creativity and diversity needed to compete in a global economy. As Malcolm Gladwell,* the author of several books on social science, wrote in 2000,"To reread *Death and Life* today is to be struck by how the intervening years have given her arguments a new and unexpected relevance. Who, after all, has a direct interest in creating diverse, vital spaces that foster creativity and serendipity? Employers do. On the fortieth anniversary of its publication, *Death and Life* has been reborn as a primer on workplace design."[4]

> *"But in some places, such as Argentina, where a Spanish translation of one of her works lay deep in the stacks of the one library that owned it, Jane Jacobs is only now getting a following, as a nascent preservation movement in creatively destructive Buenos Aires seeks a philosophic grounding... In all these places, we are reminded that Jane Jacobs remains, 50 years after Death and Life, an important figure whose influence continues to evolve."*
>
> ——Max Page and Timothy Mennel, *Reconsidering Jane Jacobs*

Future Directions

The main group of theorists and planners carrying out Jacobs's ideas and applying her prescriptions call themselves New Urbanists.*[5] In sum,"New Urbanists, who regard *Death and Life* as the most important initial theoretical contribution to their movement, not only rely on Jacobs's concepts of city density, walkable communities and 'street eyes,' but also on her belief in the mix of uses, buildings and people."[6] The New Urbanists have implemented many of Jacobs's suggestions in their plans and designs, particularly those that foster social interaction and a diversity of uses.

However, Jacobs distanced herself from this movement in a 2001 interview.[7] Some theorists and planners have criticized New Urbanism for taking Jacobs's ideas out of an urban context and applying them to communities more like suburbs or small towns:[8] "The main problem in the application of Jacobs' ideas to New Urbanism revolves around the question of whether diversity can be planned as theories in planning and urban design postulate."[9]

The irony is that whereas Jacobs criticized the urban planning of her time and advocated a more organic, spontaneous approach to cities, her ideas now find expression in new forms of planning within smaller non-urban communities. Although they may not be urban in character, they at least attempt to reflect Jacobs's ideas.

Summary

Death and Life fundamentally changed urban studies* and urban planning, and its impact continues. When it was published in 1961, a neighborhood movement in Greenwich Village, involving Jacobs herself, fought the urban renewal* projects of the architect and planner Robert Moses.* These events shaped *Death and Life*, and the movement's success validated the book's objections to urban renewal. In debunking the orthodox* theories implemented by Moses and other planners, Jacobs offered an alternative set of ideas about how cities work by fostering social interaction. Her suggestions for mixed-use* spaces, walkable neighborhoods, and historical preservation* have since been widely accepted in urban planning.

The historical significance and wide-ranging impact of *Death and Life* make it an indispensable work in urban studies and planning. It has since become a foundational text for contemporary theories and planners, such as those associated with New Urbanism. Even when scholars such as Edward Glaeser* respectfully disagree with Jacobs's ideas, *Death and Life* continues to be a major touchstone in the field.[10] Her analysis of Greenwich Village also anticipated the onset of gentrification that has transformed numerous city neighborhoods in the decades since *Death and Life* was published. Scholars including the sociologist* Sharon Zukin* have argued that Jacobs neglected the inequalities of race and class that have come to the fore with gentrification, but here again Jacobs's ability to connect cities and culture was groundbreaking.[11]

The political scientist Marshall Berman* wrote that Jacobs described the "modernism of the street" in *Death and Life*.[12] What he meant was that she captured the excitement, creativity, and dynamism generated by the mixture of people in an urban environment. Her vision represented the antithesis of that held by planners such as Moses, who sought to destroy cities because they feared the streets and dense concentrations of people. However, those fears proved unfounded, and as Jacobs's ideas gained favor, the projects championed by Moses and his ilk bit the dust before they ever got off the ground. And more than 50 years after Jacobs and Moses locked horns in a Greenwich Village showdown, cities are here to stay. As they continue to grow, so too, it appears, will Jacobs's influence.

---

1 Mark Gottdiener, Ray Hutchison, and Michael T. Ryan, *The New Urban Sociology*, 5th edn (Boulder, CO: Westview Press, 2015), 9.

2 Gottdiener, Hutchison, and Ryan, *The New Urban Sociology*, 10.

3 Richard Florida, *The Rise of the Creative Class* (New York: Basic Books, 2002).

4 Malcolm Gladwell, "Designs for Working," *The New Yorker*, December 11, 2000.

5 Peter Katz, *The New Urbanism: Towards an Architecture of Community* (New York: McGraw-Hill, 1994).

6 Matthias Wendt, "The Importance of *Death and Life of Great American Cities* by Jane Jacobs to the Profession of Urban Planning," *New Visions for Public Affairs* 1 (Spring 2009).

7 Jane Jacobs and Bill Steigerwald, "City Views: Urban Studies Legend Jane Jacobs on Gentrification, the New Urbanism, and Her Legacy," *Reason* 33, no. 2 (2001): 48–55.

8 See Max Page and Timothy Mennel, eds, *Reconsidering Jane Jacobs* (Chicago: American Planning Association, 2011).

9 Wendt, "The Importance of *Death and Life of Great American Cities*."

10 Edward Glaeser, *Triumph of the City: How Our Greatest Invention Makes Us Richer, Smarter, Greener, Healthier, and Happier* (New York: Penguin Press, 2011).

11 Sharon Zukin, *Naked City: The Death and Life of Authentic Urban Spaces* (New York: Oxford University Press, 2010).

12 Marshall Berman, *All That Is Solid Melts into Air: The Experience of Modernity* (New York: Penguin Books, 1982), 314.

GLOSSARY OF TERMS

1. ***Architectural Forum*****:** an American magazine of architecture and home design published from 1892 to 1974.
2. **Chicago School:** best known for its urban sociology, this was a faculty in the sociology department at the University of Chicago that produced groundbreaking studies of urban life during the 1920s and 1930s. Chicago sociologists undertook many ethnographic studies of urban subcultures and lifestyles, and their theories compared social life in the city to an ecological system.
3. **Cross Bronx Expressway:** a major highway for automobiles designed by Robert Moses and built between 1948 and 1972. Its construction entailed the demolition of thousands of homes and business in the southern part of the Bronx, with destructive long-term consequences for its neighborhoods.
4. **Decentrists:** a group of urban theorists and planners from the nineteenth century who sought to reform the social and environmental ills of city life by decentralizing the population and built environment of cities.

5. **Garden city:** a utopian ideal of urban planning developed by Ebenezer Howard in the late nineteenth century as an alternative to the densely populated cities of the time. Its design was supposed to combine the best features of city and country living.
6. **Gentrification:** the migration of white-collar (middle-class) professionals to formerly deteriorated urban neighborhoods. It increases property values and housing costs, displacing its previous low-income inhabitants, and incorporates the cultural tastes of white-collar workers for food, art, music, and recreation.
7. **Greenwich Village:** a neighborhood in Lower Manhattan that developed a reputation as a bohemian, free-thinking haven for writers, artists, musicians, and political dissidents.
8. **Historical preservation:** a movement to protect, preserve, and restore buildings, monuments, and objects with historic significance in urban areas. This movement was energized by community opposition to the demolition projects of urban renewal.

9. **Joint Committee to Stop the Lower Manhattan Expressway:** a local movement that succeeded in stopping the construction of an automotive expressway designed by Robert Moses to cut through the neighborhoods and business districts of Lower Manhattan.

10. **Lower Manhattan Expressway:** the plan proposed by the planner Robert Moses to build an automotive expressway that would connect the eastern and western ends of Lower Manhattan.The project would have entailed the demolition of the neighborhoods now known as SoHo and Little Italy. Community opposition forced the project's cancellation in 1962.

11. **Marxism:** both a methodology for sociological analysis and a theory of historical development. Inspired by the writings of the German economist and political philosopher Karl Marx, its traditional emphasis has been on class conflict and the economic determination of behavior, and on providing a systemic and incisive critique of the capitalist (that is, profit-oriented) economy.

12. **Mixed-use development:** districts or zones that blend a combination of industrial, commercial, residential, and cultural spaces.The benefits of this combination include reduced distances between destinations, stronger neighborhood character, and more bicycle-and pedestrian-friendly streets. Mixed-use districts are increasingly associated with gentrification.

13. **New Urbanism:** a movement in urban design and planning that arose in the 1980s and 1990s. New Urbanists have designed numerous neighborhoods that incorporate walkable streets, downtown centers, historical preservation, and environmental sustainability.

14. **Open city:** an urban ideal for a city based on principles of democracy and diversity proposed by the urban scholars Richard Burdett and Richard Sennett of the London School of Economics. Its key features are passage territories, incomplete form, narratives of development, and democratic space.

15. **Orthodox:** something that conforms to traditional or generally accepted rules or practice.

16. **Orthodoxy:** a generally accepted theory, doctrine, or practice.

17. **Postmodern architecture:** a movement starting in the late 1970s involving architects who sought to move on from the International Style that dominated postwar architecture. In contrast to architectural modernism, it incorporates self-referential wit and non-functional ornamentation.

18. **Radiant City:** the architect Le Corbusier's model of an ideal city, developed in the 1920s and 1930s.This ideal city would be composed of high-rise buildings, efficient flows of traffic, and abundant green spaces.

19. **Second-wave feminism:** the social movement of women that began in the late 1960s, continuing earlier movements of women for suffrage, nondiscrimination, reproductive rights, and social equality.

20. **Sociology:** the study of the history, formation, and structures of human societies.

21. **Spadina Expressway:** a proposed freeway that would have run through downtown Toronto.The project was canceled in 1971 as a result of public opposition.

22. ***The New Yorker*:** a weekly magazine published since 1925 that includes fiction, journalism, poetry, criticism, satire, and cartoons.

23. **Urban planning:** the technical process of designing cities with concerns for infrastructure, transportation, communications, and public welfare.

24. **Urban renewal:** projects of reconstruction that were most often implemented in cities after World War II. Urban renewal typically involved the demolition of inner-city neighborhoods for the construction of automotive highways.

25. **Urban sociology:** the study of the social constitution of urban environments.

26. **Urban studies:** an academic field that combines multiple disciplines for the study of cities and their suburbs. Its most common subtopics include urban economics, urban planning, urban politics, urban transportation, and urban sociology.

27. **Washington Square Park:** a public square and gathering place central to the neighborhood of Greenwich Village in New York City. Since the late nineteenth century, this park has been a vital spot for congregating and

performing for musicians, artists, poets, and entertainers.

28. **Washington Square Park Committee:** a group founded by Shirley Hayes in 1952 to halt New York City's plans to extend automobile traffic through Washington Square Park in Greenwich Village.

29. **World War II:** global conflict from 1939 to 1945 that involved the world's great powers and numerous other countries around the globe.

 PEOPLE MENTIONED IN THE TEXT 

1. **Jane Addams (1860–1935)** was a social reformer and the co-founder of Hull House, a social reform center in Chicago. Her social work increased awareness of the problems of poverty and public health in cities.
2. **Marshall Berman (1940–2013)** was Distinguished Professor of Political Science at the City College of New York and the Graduate Center of the City University of New York. He was best known for his analysis of modernity from a Marxist humanist perspective.
3. **Richard Burdett** is professor of urban studies at the London School of Economics. He has played a central role in developing the idea of an "open city".
4. **Ernest Burgess (1886–1966)** was an urban sociologist at the University of Chicago. He is most noted for his theory of how cities grow in concentric zones.

5. **Rachel Carson (1907–1964)** was a marine biologist and conservationist. Her book *Silent Spring* exposed the harmful effects of pesticide use and influenced the beginnings of the environmental movement.
6. **Richard Florida (b. 1957)** is an urban theorist and professor at the Rotman School of Management at the University of Toronto. He is best known for his writings about how cities can stimulate the economic innovation of a "creative class".
7. **Betty Friedan (1921–2006)** was a feminist writer as well as the co-founder and first president of the National Organization of Women (NOW). Her book *The Feminine Mystique* is often credited with being a catalyst for the development of second-wave feminism in the 1960s.
8. **Herbert Gans (b. 1927)** is a sociologist who taught at Columbia University between 1971 and 2007. His work has spanned a wide range of topics, including the effects of urban renewal, the lives of the poor, and the workings of the news media.
9. **Sir Patrick Geddes (1854–1932)** was a Scottish intellectual and a pioneering figure in urban planning. He developed ideas for regional planning that

opposed the gridiron plans that dominated cities in his time.

10. **Malcolm Gladwell (b. 1963)** is a journalist and has been a staff writer at *The New Yorker* magazine since 1996. His five books exploring the surprising implications of academic social science have all been *New York Times* bestsellers.

11. **Edward Glaeser (b. 1967)** is professor of economics at Harvard University. His research examines how cities foster economic prosperity and environmental sustainability.

12. **Michael Harrington (1928–1989)** was a writer and founder of the Democratic Socialists of America. His first book, *The Other America*, helped launch the War on Poverty in the United States during the 1960s.

13. **David Harvey (b. 1935)** is Distinguished Professor of Anthropology and Geography at the City University of New York. He is known for his Marxist analysis of the relationship between capitalism and social space.

14. **Shirley Hayes (1912–2002)** was a community organizer in New York's Greenwich Village. She founded the Washington Square Park Committee to combat Robert Moses' plans to build expressways across Lower Manhattan.

15. **Ebenezer Howard (1850–1928)** was an English theorist of urban planning. He envisioned and designed a garden city based on utopian ideas that combined the best aspects of city and country living.

16. **Le Corbusier (1887–1965)** was a Swiss French architect and urban planner. His architectural ideas informed the redesign of Paris in the 1920s and 1930s, and his buildings can be found in many parts of the world.

17. **Robert Moses (1888–1981)** was an urban planner and public official who held numerous positions in New York City between 1922 and 1968. He is known as the "master builder" who transformed New York and its surrounding suburbs in the mid-twentieth century.

18. **Lewis Mumford (1895–1990)** was an American writer on cities and the architecture critic for *The New Yorker* magazine. He is best known for his book *The City in History*, which won the National Book Award for

Nonfiction in 1962.

19. **Ralph Nader (b. 1932)** is a consumer advocate, environmental activist, and humanitarian. From 1992 to 2008, he was a candidate for president of the United States five times.

20. **Robert Park (1864–1944)** was an urban sociologist who taught at the University of Chicago from 1914 to 1933. His work examined issues of urban ecology, social disorganization, race relations, migration, and assimilation.

21. **Lloyd Rodwin (1919–1999)** was professor of urban studies at the Massachusetts Institute of Technology (MIT) and co-founder of the MIT–Harvard Joint Center for Urban Studies. He wrote 11 books and played an important role in urban planning during the 1950s and 1960s.

22. **Saskia Sassen (b. 1947)** is a professor of sociology at Columbia University. She has been a pioneering figure in studies of globalization and cities, with her books being translated into 21 languages.

23. **Richard Sennett (b. 1943)** is Emeritus Professor of Sociology at the London School of Economics. He has written numerous books about the development of cities, social class, public culture, and work in modern society.

24. **William H.Whyte (1917–1999)** was a sociologist, urbanist, and organizational analyst. He is best known for his 1956 book *The Organization Man*, which sold more than two million copies.

25. **Louis Wirth (1897–1952)** was a sociologist at the University of Chicago and a leading figure in the Chicago School. He is best known for his work on urbanism as a way of life in cities.

26. **Sharon Zukin** is professor of sociology at Brooklyn College and at the Graduate Center of the City University of New York. Her books have examined gentrification and the relationship between cities and culture, primarily in New York.

WORKS CITED

1. Alexiou, Alice Sparberg. *Jane Jacobs: Urban Visionary*. New Brunswick, NJ: Rutgers University Press, 2006.
2. Berman, Marshall. *All That Is Solid Melts into Air: The Experience of Modernity*. New York: Penguin Books, 1982.
3. Bressi, Todd W., ed. *The Seaside Debates: A Critique of the New Urbanism*. New York: Rizzoli, 2002.
4. Carson, Rachel. *Silent Spring*. Boston: Houghton Mifflin Company, 1962.
5. Congress for the New Urbanism. "Charter of the New Urbanism." In *The City Reader*, edited by Richard T. LeGates and Frederic Stout, 356–9. New York: Routledge, 2011.
6. Dreier, Peter. "Jane Jacobs' Radical Legacy." *National Housing Institute* 146 (Summer 2006). Accessed September 5, 2015. http://www.nhi.org/online/issues/146/janejacobslegacy.html.
7. Flint, Anthony. *Wrestling with Moses: How Jane Jacobs Took on New York's Master Builder and Transformed the American City*. New York: Random House, 2009.
8. Florida, Richard. *The Rise of the Creative Class*. New York: Basic Books, 2002.
9. Friedan, Betty. *The Feminine Mystique*. New York: W. W. Norton, 1963.
10. Gans, Herbert. "City Planning and Urban Realities." *Commentary* 33 (February 1962): 170–5.
11. ———. *The Urban Villagers: Group and Class in the Life of Italian Americans*. New York: Free Press of Glencoe, 1962.
12. Gladwell, Malcolm. "Designs for Working." *The New Yorker,* December 11, 2000.
13. Glaeser, Edward. *Triumph of the City: How Our Greatest Invention Makes Us Richer, Smarter, Greener, Healthier, and Happier*. New York: Penguin Press, 2011.
14. Gottdiener, Mark, Ray Hutchinson, and Michael T. Ryan. *The New Urban Sociology*. Boulder, CO: Westview Press, 2015.
15. Gratz, Roberta Brandes. *The Battle for Gotham: New York in the Shadow of Robert Moses and Jane Jacobs*. New York: Nation Books, 2010.
16. Harrington, Michael. *The Other America: Poverty in the United States*. New York: Touchstone, 1962.

17. Harvey, David. *The Condition of Postmodernity*. Cambridge, MA: Blackwell, 1990.

18. ———. "The New Urbanism and the Communitarian Trap." *Harvard Design Magazine* 1 (1997): 68–9.

19. Jacobs, Jane. "Downtown Is for People," *Fortune* 57 (April 1958): 157–84. Accessed September 6, 2015. http://fortune.com/2011/09/18/downtown-is-for-people-fortune-classic-1958/.

20. ———. *The Economy of Cities*. New York: Random House, 1969.

21. ———. *The Question of Separatism: Quebec and the Struggle Over Sovereignty*. New York: Random House, 1980.

22. ———. *Cities and the Wealth of Nations*. New York: Vintage Books, 1984.

23. ———. *The Death and Life of Great American Cities*. New York: Vintage Books, 1992.

24. ———. *Systems of Survival*. New York: Random House, 1992.

25. ———. *The Nature of Economies*. New York: Random House, 2000.

26. ———. *Dark Age Ahead*. New York: Random House, 2004.

27. Jacobs, Jane, and Bill Steigerwald, "City Views: Urban Studies Legend Jane Jacobs on Gentrification, the New Urbanism, and Her Legacy," *Reason* 33, no. 2 (June 2001). Accessed September 13, 2015. https://reason.com/archives/2001/06/01/city-views.

28. Katz, Peter. *The New Urbanism: Toward an Architecture of Community*. New York: McGraw-Hill, 1994.

29. Kidd, Kenneth. "Did Jane Jacobs's Critics Have a Point After All?" *The Toronto Star*, November 25, 2011. Accessed August 31, 2015. http://www.thestar. com/news/insight/2011/11/25/did_jane_jacobs_critics_have_a_point_after_ all.html.

30. LeGates, Richard T., and Frederic Stout, eds. *The City Reader*. New York: Routledge, 2011.

31. Lin, Janet, and Christopher Mele, eds. *The Urban Sociology Reader*. New York: Routledge, 2012.

32. Macionis, John J., and Vincent M. Parrillo, *Cities and Urban Life*. Boston: Pearson, 2013.

33. Mumford, Lewis. "The Sky Line: 'Mother Jacobs Home Remedies.'" *The New*

Yorker, December 1, 1962.

34. Nader, Ralph. *Unsafe at Any Speed: The Designed-In Dangers of the American Automobile*. New York: Grossman, 1965.
35. Osman, Suleiman. *The Invention of Brownstone Brooklyn: Gentrification and the Search for Authenticity in Postwar New York*. New York: Oxford University Press, 2011.
36. Page, Max, and Timothy Mennel, eds. *Reconsidering Jane Jacobs*. Chicago: American Planning Association, 2011.
37. Rodwin, Lloyd. "Neighbors Are Needed." *New York Times*, November 5, 1961.
38. Sassen, Saskia. "What Would Jane Jacobs See in the Global City? Place and Social Practices." In *The Urban Wisdom of Jane Jacobs*, edited by Sonia Hirt and Diana Zahm, 84–99. New York: Routledge, 2012.
39. Sennett, Richard. *Flesh and Stone: The Body and the City in Western Civilization*. New York: W. W. Norton, 1994.
40. ———. "The Open City." *Urban Age* (November 2006): 2–3.
41. Strausbaugh, John. *The Village: A History of Greenwich Village*. New York: Harper Collins, 2013.
42. Ward, Stephen. "Obituary: Jane Jacobs." *The Independent*, June 3, 2006. Accessed August 29, 2015. http://www.independent.co.uk/news/obituaries/ jane-jacobs-6099183.html.
43. Wendt, Matthias. "The Importance of *Death and Life of Great American Cities* by Jane Jacobs to the Profession of Urban Planning," *New Visions for Public Affairs* 1 (Spring 2009). Accessed September 13, 2015. https://nvpajournal.wordpress.com/issues/volume-1/.
44. Wirth, Louis. "Urbanism as a Way of Life." *American Journal of Sociology* 44, no. 1 (July 1938): 1–24.
45. Zukin, Sharon. *Landscapes of Power: From Detroit to Disney World*. Berkeley and Los Angeles: University of California Press, 1991.
46. ———. "Changing Landscapes of Power: Opulence and the Urge for Authenticity." *International Journal of Urban and Regional Research* 33, no. 2 (2009): 548–9.
47. ———. *Naked City: The Death and Life of Authentic Urban Places*. New York: Oxford University Press, 2010.

原书作者简介

简·雅各布斯原名简·布茨纳，美国作家、记者、社会活动家，1916年生于宾夕法尼亚州斯克兰顿市。她于1934年移居纽约，并成为一名记者，为《建筑论坛》和《财富》等杂志撰稿。20世纪50年代末，作为下曼哈顿格林尼治村的一位居民，她参加了一场草根运动，挽救了所在社区为腾出空间建造规划中的新高速公路而被拆除的命运。在1961年出版的《美国大城市的死与生》一书中，雅各布斯表达了她对占据统治地位却考虑欠周的城市规划思想与政策的反对。自那以后，许多城市规划师都接纳了她的思想，让城市保持多样性、适合步行和紧密集中。雅各布斯2006年去世，享年89岁。

本书作者简介

马丁·富勒获剑桥大学社会学博士学位，在纽约和柏林专事艺术社会学研究，目前是柏林工业大学研究员。

瑞恩·摩尔获加利福尼亚大学圣地亚哥分校社会学与文化分析博士学位，曾在美国多所大学从事教学工作，著有《像少年英气一样销售：音乐、青年文化与社会危机》（纽约：纽约大学出版社，2009年）。

世界名著中的批判性思维

《世界思想宝库钥匙丛书》致力于深入浅出地阐释全世界著名思想家的观点，不论是谁、在何处都能了解到，从而推进批判性思维发展。

《世界思想宝库钥匙丛书》与世界顶尖大学的一流学者合作，为一系列学科中最有影响的著作推出新的分析文本，介绍其观点和影响。在这一不断扩展的系列中，每种选入的著作都代表了历经时间考验的思想典范。通过为这些著作提供必要背景、揭示原作者的学术渊源以及说明这些著作所产生的影响，本系列图书希望让读者以新视角看待这些划时代的经典之作。读者应学会思考、运用并挑战这些著作中的观点，而不是简单接受它们。

ABOUT THE AUTHOR OF THE ORIGINAL WORK

American author, journalist, and activist **Jane Jacobs** was born Jane Butzner in 1916 in Scranton, Pennsylvania. She moved to New York City in 1934, where she became a journalist, writing for magazines including Architectural Forum and Fortune. As a resident of Lower Manhattan's Greenwich Village, she joined a grassroots movement in the late 1950s to save her neighborhood from its planned destruction to make way for new expressways. Jacobs expressed her opposition to dominant yet ill-conceived ideas of city planning and policy in her 1961 work *The Death and Life of Great American Cities.* Many urban planners have since adopted its ideas to make cities more diverse, walkable, and densely concentrated. Jacobs died in 2006 at the age of 89.

ABOUT THE AUTHORS OF THE ANALYSIS

Dr. Martin Fuller holds a PhD in sociology from the University of Cambridge, focusing on the sociology of art in New York and Berlin. He is currently a researcher at the Technische Universität, Berlin.

Dr. Ryan Moore holds PhDs in sociology and cultural analysis from the University of California, San Diego. He has taught at universities across America and is the author of *Sells Like Teen Spirit: Music, Youth Culture, and Social Crisis* (New York: NYU Press, 2009).

ABOUT MACAT
GREAT WORKS FOR CRITICAL THINKING

Macat is focused on making the ideas of the world's great thinkers accessible and comprehensible to everybody, everywhere, in ways that promote the development of enhanced critical thinking skills.

It works with leading academics from the world's top universities to produce new analyses that focus on the ideas and the impact of the most influential works ever written across a wide variety of academic disciplines. Each of the works that sit at the heart of its growing library is an enduring example of great thinking. But by setting them in context — and looking at the influences that shaped their authors, as well as the responses they provoked — Macat encourages readers to look at these classics and game-changers with fresh eyes. Readers learn to think, engage and challenge their ideas, rather than simply accepting them.

批判性思维与《美国大城市的死与生》

首要批判性思维技巧：评估

次要批判性思维技巧：理性化思维

尽管没有受过正规的城市规划训练，简·雅各布斯却娴熟地把握了第二次世界大战后的年代里美国城市规划者提出的政策中的长处与弱点。规划者们认为，在美国城市发展中，汽车的有效行动要比生活于都市中的人们的日常生活更重要。

在 1961 年出版的著作《美国大城市的死与生》中，雅各布斯细致研究了这些观点的相关性，通过曝光它们的短视之处，祛除了这些观点。她评估了手上的有关信息，得出大相径庭的结论，即城市规划者毁灭了大城市，因为他们不明白令城市伟大的是城市中的社会交往活动。由规划理论推演出的建议和政策没考虑到城市生活的社会动力学元素。他们拘泥于现代生活方式虚幻的前卫模式，与现实没有任何联系，也与真正生活在这些空间中的人们的真正欲求没有联系。专家们四处游说分隔与标准化，将商业、居住、工业和文化空间分隔开。但城市规划一个真正具有前瞻性的路径是应该聚合混合的用途空间，连同适合步行的短街区、高度集中的人群和新旧搭配的建筑物一起，这才会创造出真正的城市活力。

CRITICAL THINKING AND *THE DEATH AND LIFE OF GREAT AMERICAN CITIES*

- Primary critical thinking skill: EVALUATION
- Secondary critical thinking skill: REASONING

Despite having no formal training in urban planning, Jane Jacobs deftly explores the strengths and weaknesses of policy arguments put forward by American urban planners in the era after World War II.They believed that the efficient movement of cars was of more value in the development of US cities than the everyday lives of the people living there.

By carefully examining their relevance in her 1961 book, *The Death and Life of Great American Cities*, Jacobs dismantles these arguments by highlighting their shortsightedness. She evaluates the information to hand and comes to a very different conclusion, that urban planners ruin great cities, because they don't understand that it is a city's social interaction that makes it great. Proposals and policies that are drawn from planning theory do not consider the social dynamics of city life.They are in thrall to futuristic fantasies of a modern way of living that bears no relation to reality, or to the desires of real people living in real spaces. Professionals lobby for separation and standardization, splitting commercial, residential, industrial, and cultural spaces. However, a truly visionary approach to urban planning should incorporate spaces with mixed uses, together with short, walkable blocks, large concentrations of people, and a mix of new and old buildings. This creates true urban vitality.

《世界思想宝库钥匙丛书》简介

《世界思想宝库钥匙丛书》致力于为一系列在各领域产生重大影响的人文社科类经典著作提供独特的学术探讨。每一本读物都不仅仅是原经典著作的内容摘要，而是介绍并深入研究原经典著作的学术渊源、主要观点和历史影响。这一丛书的目的是提供一套学习资料，以促进读者掌握批判性思维，从而更全面、深刻地去理解重要思想。

每一本读物分为 3 个部分：学术渊源、学术思想和学术影响，每个部分下有 4 个小节。这些章节旨在从各个方面研究原经典著作及其反响。

由于独特的体例，每一本读物不但易于阅读，而且另有一项优点：所有读物的编排体例相同，读者在进行某个知识层面的调查或研究时可交叉参阅多本该丛书中的相关读物，从而开启跨领域研究的路径。

为了方便阅读，每本读物最后还列出了术语表和人名表（在书中则以星号 * 标记），此外还有参考文献。

《世界思想宝库钥匙丛书》与剑桥大学合作，理清了批判性思维的要点，即如何通过 6 种技能来进行有效思考。其中 3 种技能让我们能够理解问题，另 3 种技能让我们有能力解决问题。这 6 种技能合称为“批判性思维 PACIER 模式”，它们是：

分析：了解如何建立一个观点；
评估：研究一个观点的优点和缺点；
阐释：对意义所产生的问题加以理解；
创造性思维：提出新的见解，发现新的联系；
解决问题：提出切实有效的解决办法；
理性化思维：创建有说服力的观点。

了解更多信息，请浏览 www.macat.com。

THE MACAT LIBRARY

The Macat Library is a series of unique academic explorations of seminal works in the humanities and social sciences — books and papers that have had a significant and widely recognised impact on their disciplines. It has been created to serve as much more than just a summary of what lies between the covers of a great book. It illuminates and explores the influences on, ideas of, and impact of that book. Our goal is to offer a learning resource that encourages critical thinking and fosters a better, deeper understanding of important ideas.

Each publication is divided into three Sections: Influences, Ideas, and Impact. Each Section has four Modules. These explore every important facet of the work, and the responses to it.

This Section-Module structure makes a Macat Library book easy to use, but it has another important feature. Because each Macat book is written to the same format, it is possible (and encouraged!) to cross-reference multiple Macat books along the same lines of inquiry or research. This allows the reader to open up interesting interdisciplinary pathways.

To further aid your reading, lists of glossary terms and people mentioned are included at the end of this book (these are indicated by an asterisk [*] throughout) — as well as a list of works cited.

Macat has worked with the University of Cambridge to identify the elements of critical thinking and understand the ways in which six different skills combine to enable effective thinking.

Three allow us to fully understand a problem; three more give us the tools to solve it. Together, these six skills make up the PACIER model of critical thinking. They are:

ANALYSIS — understanding how an argument is built
EVALUATION — exploring the strengths and weaknesses of an argument
INTERPRETATION — understanding issues of meaning
CREATIVE THINKING — coming up with new ideas and fresh connections
PROBLEM-SOLVING — producing strong solutions
REASONING — creating strong arguments

To find out more, visit WWW.MACAT.COM.

“《世界思想宝库钥匙丛书》提供了独一无二的跨学科学习和研究工具。它介绍那些革新了各自学科研究的经典著作，还邀请全世界一流专家和教育机构进行严谨的分析，为每位读者打开世界顶级教育的大门。”

——安德烈亚斯·施莱歇尔，
经济合作与发展组织教育与技能司司长

“《世界思想宝库钥匙丛书》直面大学教育的巨大挑战……他们组建了一支精干而活跃的学者队伍，来推出在研究广度上颇具新意的教学材料。”

——布罗尔斯教授、勋爵，剑桥大学前校长

“《世界思想宝库钥匙丛书》的愿景令人赞叹。它通过分析和阐释那些曾深刻影响人类思想以及社会、经济发展的经典文本，提供了新的学习方法。它推动批判性思维，这对于任何社会和经济体来说都是至关重要的。这就是未来的学习方法。”

——查尔斯·克拉克阁下，英国前教育大臣

“对于那些影响了各自领域的著作，《世界思想宝库钥匙丛书》能让人们立即了解到围绕那些著作展开的评论性言论，这让该系列图书成为在这些领域从事研究的师生们不可或缺的资源。”

——威廉·特朗佐教授，加利福尼亚大学圣地亚哥分校

"Macat offers an amazing first-of-its-kind tool for interdisciplinary learning and research. Its focus on works that transformed their disciplines and its rigorous approach, drawing on the world's leading experts and educational institutions, opens up a world-class education to anyone."

—— Andreas Schleicher, Director for Education and Skills, Organisation for Economic Co-operation and Development

"Macat is taking on some of the major challenges in university education... They have drawn together a strong team of active academics who are producing teaching materials that are novel in the breadth of their approach."

—— Prof Lord Broers, former Vice-Chancellor of the University of Cambridge

"The Macat vision is exceptionally exciting. It focuses upon new modes of learning which analyse and explain seminal texts which have profoundly influenced world thinking and so social and economic development. It promotes the kind of critical thinking which is essential for any society and economy. This is the learning of the future."

—— Rt Hon Charles Clarke, former UK Secretary of State for Education

"The Macat analyses provide immediate access to the critical conversation surrounding the books that have shaped their respective discipline, which will make them an invaluable resource to all of those, students and teachers, working in the field."

—— Prof William Tronzo, University of California at San Diego

TITLE	中文书名	类别
An Analysis of Arjun Appadurai's *Modernity at Large: Cultural Dimensions of Globalisation*	解析阿尔君·阿帕杜莱《消失的现代性：全球化的文化维度》	人类学
An Analysis of Claude Lévi-Strauss's *Structural Anthropology*	解析克劳德·列维–斯特劳斯《结构人类学》	人类学
An Analysis of Marcel Mauss's *The Gift*	解析马塞尔·莫斯《礼物》	人类学
An Analysis of Jared M. Diamond's *Guns, Germs, and Steel: The Fate of Human Societies*	解析贾雷德·戴蒙德《枪炮、病菌与钢铁：人类社会的命运》	人类学
An Analysis of Clifford Geertz's *The Interpretation of Cultures*	解析克利福德·格尔茨《文化的解释》	人类学
An Analysis of Philippe Ariès's *Centuries of Childhood: A Social History of Family Life*	解析菲力浦·阿利埃斯《儿童的世纪：旧制度下的儿童和家庭生活》	人类学
An Analysis of W. Chan Kim & Renée Mauborgne's *Blue Ocean Strategy*	解析金伟灿 / 勒妮·莫博涅《蓝海战略》	商业
An Analysis of John P. Kotter's *Leading Change*	解析约翰·P. 科特《领导变革》	商业
An Analysis of Michael E. Porter's *Competitive Strategy: Creating and Sustaining Superior Performance*	解析迈克尔·E. 波特《竞争战略：分析产业和竞争对手的技术》	商业
An Analysis of Jean Lave & Etienne Wenger's *Situated Learning: Legitimate Peripheral Participation*	解析琼·莱夫 / 艾蒂纳·温格《情境学习：合法的边缘性参与》	商业
An Analysis of Douglas McGregor's *The Human Side of Enterprise*	解析道格拉斯·麦格雷戈《企业的人性面》	商业
An Analysis of Milton Friedman's *Capitalism and Freedom*	解析米尔顿·弗里德曼《资本主义与自由》	商业
An Analysis of Ludwig von Mises's *The Theory of Money and Credit*	解析路德维希·冯·米塞斯《货币和信用理论》	经济学
An Analysis of Adam Smith's *The Wealth of Nations*	解析亚当·斯密《国富论》	经济学
An Analysis of Thomas Piketty's *Capital in the Twenty-First Century*	解析托马斯·皮凯蒂《21 世纪资本论》	经济学
An Analysis of Nassim Nicholas Taleb's *The Black Swan: The Impact of the Highly Improbable*	解析纳西姆·尼古拉斯·塔勒布《黑天鹅：如何应对不可预知的未来》	经济学
An Analysis of Ha-Joon Chang's *Kicking Away the Ladder*	解析张夏准《富国陷阱：发达国家为何踢开梯子》	经济学
An Analysis of Thomas Robert Malthus's *An Essay on the Principle of Population*	解析托马斯·马尔萨斯《人口论》	经济学

An Analysis of John Maynard Keynes's *The General Theory of Employment, Interest and Money*	解析约翰·梅纳德·凯恩斯《就业、利息和货币通论》	经济学
An Analysis of Milton Friedman's *The Role of Monetary Policy*	解析米尔顿·弗里德曼《货币政策的作用》	经济学
An Analysis of Burton G. Malkiel's *A Random Walk Down Wall Street*	解析伯顿·G. 马尔基尔《漫步华尔街》	经济学
An Analysis of Friedrich A. Hayek's *The Road to Serfdom*	解析弗里德里希·A. 哈耶克《通往奴役之路》	经济学
An Analysis of Charles P. Kindleberger's *Manias, Panics, and Crashes: A History of Financial Crises*	解析查尔斯·P. 金德尔伯格《疯狂、惊恐和崩溃：金融危机史》	经济学
An Analysis of Amartya Sen's *Development as Freedom*	解析阿马蒂亚·森《以自由看待发展》	经济学
An Analysis of Rachel Carson's *Silent Spring*	解析蕾切尔·卡森《寂静的春天》	地理学
An Analysis of Charles Darwin's *On the Origin of Species: by Means of Natural Selection, or The Preservation of Favoured Races in the Struggle for Life*	解析查尔斯·达尔文《物种起源》	地理学
An Analysis of World Commission on Environment and Development's *The Brundtland Report, Our Common Future*	解析世界环境与发展委员会《布伦特兰报告：我们共同的未来》	地理学
An Analysis of James E. Lovelock's *Gaia: A New Look at Life on Earth*	解析詹姆斯·E. 拉伍洛克《盖娅：地球生命的新视野》	地理学
An Analysis of Paul Kennedy's *The Rise and Fall of the Great Powers: Economic Change and Military Conflict from 1500—2000*	解析保罗·肯尼迪《大国的兴衰：1500—2000 年的经济变革与军事冲突》	历史
An Analysis of Janet L. Abu-Lughod's *Before European Hegemony: The World System A. D. 1250—1350*	解析珍妮特·L. 阿布–卢格霍德《欧洲霸权之前：1250—1350 年的世界体系》	历史
An Analysis of Alfred W. Crosby's *The Columbian Exchange: Biological and Cultural Consequences of 1492*	解析艾尔弗雷德·W. 克罗斯比《哥伦布大交换：1492 年以后的生物影响和文化冲击》	历史
An Analysis of Tony Judt's *Postwar: A History of Europe since 1945*	解析托尼·贾德《战后欧洲史》	历史
An Analysis of Richard J. Evans's *In Defence of History*	解析理查德·J. 艾文斯《捍卫历史》	历史
An Analysis of Eric Hobsbawm's *The Age of Revolution: Europe 1789–1848*	解析艾瑞克·霍布斯鲍姆《革命的年代：欧洲 1789—1848 年》	历史

An Analysis of Roland Barthes's *Mythologies*	解析罗兰·巴特《神话学》	文学与批判理论
An Analysis of Simon de Beauvoir's *The Second Sex*	解析西蒙娜·德·波伏娃《第二性》	文学与批判理论
An Analysis of Edward W. Said's *Orientalism*	解析爱德华·W. 萨义德《东方主义》	文学与批判理论
An Analysis of Virginia Woolf's *A Room of One's Own*	解析弗吉尼亚·伍尔芙《一间自己的房间》	文学与批判理论
An Analysis of Judith Butler's *Gender Trouble*	解析朱迪斯·巴特勒《性别麻烦》	文学与批判理论
An Analysis of Ferdinand de Saussure's *Course in General Linguistics*	解析费尔迪南·德·索绪尔《普通语言学教程》	文学与批判理论
An Analysis of Susan Sontag's *On Photography*	解析苏珊·桑塔格《论摄影》	文学与批判理论
An Analysis of Walter Benjamin's *The Work of Art in the Age of Mechanical Reproduction*	解析瓦尔特·本雅明《机械复制时代的艺术作品》	文学与批判理论
An Analysis of W.E.B. Du Bois's *The Souls of Black Folk*	解析 W.E.B. 杜博伊斯《黑人的灵魂》	文学与批判理论
An Analysis of Plato's *The Republic*	解析柏拉图《理想国》	哲学
An Analysis of Plato's *Symposium*	解析柏拉图《会饮篇》	哲学
An Analysis of Aristotle's *Metaphysics*	解析亚里士多德《形而上学》	哲学
An Analysis of Aristotle's *Nicomachean Ethics*	解析亚里士多德《尼各马可伦理学》	哲学
An Analysis of Immanuel Kant's *Critique of Pure Reason*	解析伊曼努尔·康德《纯粹理性批判》	哲学
An Analysis of Ludwig Wittgenstein's *Philosophical Investigations*	解析路德维希·维特根斯坦《哲学研究》	哲学
An Analysis of G.W.F. Hegel's *Phenomenology of Spirit*	解析 G.W.F. 黑格尔《精神现象学》	哲学
An Analysis of Baruch Spinoza's *Ethics*	解析巴鲁赫·斯宾诺莎《伦理学》	哲学
An Analysis of Hannah Arendt's *The Human Condition*	解析汉娜·阿伦特《人的境况》	哲学
An Analysis of G.E.M. Anscombe's *Modern Moral Philosophy*	解析 G.E.M. 安斯康姆《现代道德哲学》	哲学
An Analysis of David Hume's *An Enquiry Concerning Human Understanding*	解析大卫·休谟《人类理解研究》	哲学

An Analysis of Søren Kierkegaard's *Fear and Trembling*	解析索伦·克尔凯郭尔《恐惧与战栗》	哲学
An Analysis of René Descartes's *Meditations on First Philosophy*	解析勒内·笛卡尔《第一哲学沉思录》	哲学
An Analysis of Friedrich Nietzsche's *On the Genealogy of Morality*	解析弗里德里希·尼采《论道德的谱系》	哲学
An Analysis of Gilbert Ryle's *The Concept of Mind*	解析吉尔伯特·赖尔《心的概念》	哲学
An Analysis of Thomas Kuhn's *The Structure of Scientific Revolutions*	解析托马斯·库恩《科学革命的结构》	哲学
An Analysis of John Stuart Mill's *Utilitarianism*	解析约翰·斯图亚特·穆勒《功利主义》	哲学
An Analysis of Aristotle's *Politics*	解析亚里士多德《政治学》	政治学
An Analysis of Niccolò Machiavelli's *The Prince*	解析尼科洛·马基雅维利《君主论》	政治学
An Analysis of Karl Marx's *Capital*	解析卡尔·马克思《资本论》	政治学
An Analysis of Benedict Anderson's *Imagined Communities*	解析本尼迪克特·安德森《想象的共同体》	政治学
An Analysis of Samuel P. Huntington's *The Clash of Civilizations and the Remaking of World Order*	解析塞缪尔·P. 亨廷顿《文明的冲突与世界秩序重建》	政治学
An Analysis of Alexis de Tocqueville's *Democracy in America*	解析阿列克西·德·托克维尔《论美国的民主》	政治学
An Analysis of J. A. Hobson's *Imperialism: A Study*	解析约·阿·霍布森《帝国主义》	政治学
An Analysis of Thomas Paine's *Common Sense*	解析托马斯·潘恩《常识》	政治学
An Analysis of John Rawls's *A Theory of Justice*	解析约翰·罗尔斯《正义论》	政治学
An Analysis of Francis Fukuyama's *The End of History and the Last Man*	解析弗朗西斯·福山《历史的终结与最后的人》	政治学
An Analysis of John Locke's *Two Treatises of Government*	解析约翰·洛克《政府论》	政治学
An Analysis of Sun Tzu's *The Art of War*	解析孙武《孙子兵法》	政治学
An Analysis of Henry Kissinger's *World Order: Reflections on the Character of Nations and the Course of History*	解析亨利·基辛格《世界秩序》	政治学
An Analysis of Jean-Jacques Rousseau's *The Social Contract*	解析让-雅克·卢梭《社会契约论》	政治学

An Analysis of Odd Arne Westad's *The Global Cold War: Third World Interventions and the Making of Our Times*	解析文安立《全球冷战：美苏对第三世界的干涉与当代世界的形成》	政治学
An Analysis of Sigmund Freud's *The Interpretation of Dreams*	解析西格蒙德·弗洛伊德《梦的解析》	心理学
An Analysis of William James' *The Principles of Psychology*	解析威廉·詹姆斯《心理学原理》	心理学
An Analysis of Philip Zimbardo's *The Lucifer Effect*	解析菲利普·津巴多《路西法效应》	心理学
An Analysis of Leon Festinger's *A Theory of Cognitive Dissonance*	解析利昂·费斯汀格《认知失调论》	心理学
An Analysis of Richard H. Thaler & Cass R. Sunstein's *Nudge: Improving Decisions about Health, Wealth, and Happiness*	解析理查德·H. 泰勒 / 卡斯·R. 桑斯坦《助推：如何做出有关健康、财富和幸福的更优决策》	心理学
An Analysis of Gordon Allport's *The Nature of Prejudice*	解析高尔登·奥尔波特《偏见的本质》	心理学
An Analysis of Steven Pinker's *The Better Angels of Our Nature: Why Violence Has Declined*	解析斯蒂芬·平克《人性中的善良天使：暴力为什么会减少》	心理学
An Analysis of Stanley Milgram's *Obedience to Authority*	解析斯坦利·米尔格拉姆《对权威的服从》	心理学
An Analysis of Betty Friedan's *The Feminine Mystique*	解析贝蒂·弗里丹《女性的奥秘》	心理学
An Analysis of David Riesman's *The Lonely Crowd: A Study of the Changing American Character*	解析大卫·理斯曼《孤独的人群：美国人社会性格演变之研究》	社会学
An Analysis of Franz Boas's *Race, Language and Culture*	解析弗朗兹·博厄斯《种族、语言与文化》	社会学
An Analysis of Pierre Bourdieu's *Outline of a Theory of Practice*	解析皮埃尔·布尔迪厄《实践理论大纲》	社会学
An Analysis of Max Weber's *The Protestant Ethic and the Spirit of Capitalism*	解析马克斯·韦伯《新教伦理与资本主义精神》	社会学
An Analysis of Jane Jacobs's *The Death and Life of Great American Cities*	解析简·雅各布斯《美国大城市的死与生》	社会学
An Analysis of C. Wright Mills's *The Sociological Imagination*	解析 C. 赖特·米尔斯《社会学的想象力》	社会学
An Analysis of Robert E. Lucas Jr.'s *Why Doesn't Capital Flow from Rich to Poor Countries?*	解析小罗伯特·E. 卢卡斯《为何资本不从富国流向穷国？》	社会学

An Analysis of Émile Durkheim's *On Suicide*	解析埃米尔·迪尔凯姆《自杀论》	社会学
An Analysis of Eric Hoffer's *The True Believer: Thoughts on the Nature of Mass Movements*	解析埃里克·霍弗《狂热分子：群众运动圣经》	社会学
An Analysis of Jared M. Diamond's *Collapse: How Societies Choose to Fail or Survive*	解析贾雷德·M. 戴蒙德《大崩溃：社会如何选择兴亡》	社会学
An Analysis of Michel Foucault's *The History of Sexuality Vol. 1: The Will to Knowledge*	解析米歇尔·福柯《性史（第一卷）：求知意志》	社会学
An Analysis of Michel Foucault's *Discipline and Punish*	解析米歇尔·福柯《规训与惩罚》	社会学
An Analysis of Richard Dawkins's *The Selfish Gene*	解析理查德·道金斯《自私的基因》	社会学
An Analysis of Antonio Gramsci's *Prison Notebooks*	解析安东尼奥·葛兰西《狱中札记》	社会学
An Analysis of Augustine's *Confessions*	解析奥古斯丁《忏悔录》	神学
An Analysis of C. S. Lewis's *The Abolition of Man*	解析 C.S. 路易斯《人之废》	神学

图书在版编目（CIP）数据

解析简·雅各布斯《美国大城市的死与生》：汉、英／马丁·富勒（Martin Fuller），瑞恩·摩尔（Ryan Moore）著；王青松译.
—上海：上海外语教育出版社，2019
（世界思想宝库钥匙丛书）
ISBN 978-7-5446-5840-9

Ⅰ.①解… Ⅱ.①马… ②瑞… ③王… Ⅲ.①城市规划—研究—美国—汉、英
Ⅳ.①TU984.712

中国版本图书馆CIP数据核字（2019）第076643号

出版发行：上海外语教育出版社
（上海外国语大学内） 邮编：200083
电　　话：021-65425300（总机）
电子邮箱：bookinfo@sflep.com.cn
网　　址：http://www.sflep.com
责任编辑：杨莹雪

印　　刷：上海信老印刷厂
开　　本：890×1240 1/32 印张 5.625 字数 116千字
版　　次：2019年8月第1版 2019年8月第1次印刷
印　　数：2 100册

书　　号：ISBN 978-7-5446-5840-9 / G
定　　价：30.00元
本版图书如有印装质量问题，可向本社调换
质量服务热线：4008-213-263 电子邮箱：editorial@sflep.com